CIVIL ENGINEERING FORMULAS

CIVIL ENGINEERING FORMULAS

Tyler G. Hicks, P.E.

International Engineering Associates
Member: American Society of Mechanical Engineers
United States Naval Institute

McGRAW-HILL

New York Chicago San Francisco Lisbon London
Madrid Mexico City Milan New Delhi San Juan
Seoul Singapore Sydney Toronto

Library of Congress Cataloging-in-Publication Data

Hicks, Tyler Gregory, date.

Civil engineering formulas/Tyler G. Hicks.
p. cm.
ISBN 0–07–135612–6
1. Civil engineering—Mathematics. 2. Mathematics—Formulae. I. Title.

TA331.H53 2001
624′.01′51—dc21 2001034309

McGraw-Hill

A Division of The McGraw·Hill Companies

1 2 3 4 5 6 7 8 9 0 WCL/WCL 0 7 6 5 4 3 2 1

ISBN 0–07–135612–6

The sponsoring editor for this book was Larry S. Hager, the editing supervisor was David E. Fogarty, and the production supervisor was Sherri Souffrance. It was set in Times Roman by Progressive Information Technologies, Inc.

Printed and bound by Webcom Limited.

Printed on acid-free paper.

CONTENTS

Preface xiii
Acknowledgments xv
How to Use This Book xvii

Chapter 1. Conversion Factors for Civil Engineering Practice 1

Chapter 2. Beam Formulas 15

Continuous Beams / *16*
Ultimate Strength of Continuous Beams / *53*
Beams of Uniform Strength / *63*
Safe Loads for Beams of Various Types / *64*
Rolling and Moving Loads / *79*
Curved Beams / *82*
Elastic Lateral Buckling of Beams / *88*
Combined Axial and Bending Loads / *92*
Unsymmetrical Bending / *93*
Eccentric Loading / *94*
Natural Circular Frequencies and Natural Periods of Vibration of Prismatic Beams / *96*

Chapter 3. Column Formulas 99

General Considerations / *100*
Short Columns / *102*
Eccentric Loads on Columns / *102*
Column Base Plate Design / *111*
American Institute of Steel Construction Allowable-Stress Design Approach / *113*
Composite Columns / *115*
Elastic Flexural Buckling of Columns / *118*
Allowable Design Loads for Aluminum Columns / *121*
Ultimate-Strength Design of Concrete Columns / *124*

Chapter 4. Piles and Piling Formulas 131

Allowable Loads on Piles / *132*
Laterally Loaded Vertical Piles / *133*
Toe Capacity Load / *134*
Groups of Piles / *136*
Foundation-Stability Analysis / *139*
Axial-Load Capacity of Single Piles / *143*
Shaft Settlement / *144*
Shaft Resistance to Cohesionless Soil / *145*

Chapter 5. Concrete Formulas 147

Reinforced Concrete / *148*
Water/Cementitious Materials Ratio / *148*
Job Mix Concrete Volume / *149*
Modulus of Elasticity of Concrete / *150*
Tensile Strength of Concrete / *151*
Reinforcing Steel / *151*
Continuous Beams and One-Way Slabs / *151*
Design Methods for Beams, Columns, and Other Members / *153*
Properties in the Hardened State / *167*

Tension Development Lengths / *169*
Compression Development Lengths / *170*
Crack Control of Flexural Members / *170*
Required Strength / *171*
Deflection Computations and Criteria for Concrete Beams / *172*
Ultimate-Strength Design of Rectangular Beams with Tension Reinforcement Only / *173*
Working-Stress Design of Rectangular Beams with Tension Reinforcement Only / *178*
Ultimate-Strength Design of Rectangular Beams with Compression Bars / *182*
Working-Stress Design of Rectangular Beams with Compression Bars / *183*
Ultimate-Strength Design of I and T Beams / *186*
Working-Stress Design of I and T Beams / *187*
Ultimate-Strength Design for Torsion / *189*
Working-Stress Design for Torsion / *191*
Flat-Slab Construction / *192*
Flat-Plate Construction / *195*
Shear in Slabs / *197*
Column Moments / *199*
Spirals / *200*
Braced and Unbraced Frames / *201*
Load-Bearing Walls / *202*
Shear Walls / *203*
Concrete Gravity Retaining Walls / *205*
Cantilever Retaining Walls / *208*
Wall Footings / *211*

Chapter 6. Timber Engineering Formulas 213

Grading of Lumber / *214*
Size of Lumber / *214*
Bearing / *216*
Beams / *216*
Columns / *218*
Combined Bending and Axial Load / *220*

Compression at Angle to Grain / *220*
Recommendations of the Forest Products Laboratory / *221*
Compression on Oblique Plane / *223*
Adjustments Factors for Design Values / *224*
Fasteners for Wood / *233*
Adjustment of Design Values for Connections with Fasteners / *236*
Roof Slope to Prevent Ponding / *238*
Bending and Axial Tension / *239*
Bending and Axial Compression / *240*

Chapter 7. Surveying Formulas 243

Units of Measurement / *244*
Theory of Errors / *245*
Measurement of Distance with Tapes / *247*
Vertical Control / *253*
Stadia Surveying / *253*
Photogrammetry / *255*

Chapter 8. Soil and Earthwork Formulas 257

Physical Properties of Soils / *258*
Index Parameters for Soils / *259*
Relationship of Weights and Volumes in Soils / *261*
Internal Friction and Cohesion / *263*
Vertical Pressures in Soils / *264*
Lateral Pressures in Soils, Forces on Retaining Walls / *265*
Lateral Pressure of Cohesionless Soils / *266*
Lateral Pressure of Cohesive Soils / *267*
Water Pressure / *268*
Lateral Pressure from Surcharge / *268*
Stability of Slopes / *269*
Bearing Capacity of Soils / *270*
Settlement under Foundations / *271*
Soil Compaction Tests / *272*

Compaction Equipment / 275
Formulas for Earthmoving / 276
Scraper Production / 278
Vibration Control in Blasting / 280

Chapter 9. Building and Structures Formulas 283

Load-and-Resistance Factor Design for Shear in Buildings / 284
Allowable-Stress Design for Building Columns / 285
Load-and-Resistance Factor Design for Building Columns / 287
Allowable-Stress Design for Building Beams / 287
Load-and-Resistance Factor Design for Building Beams / 290
Allowable-Stress Design for Shear in Buildings / 295
Stresses in Thin Shells / 297
Bearing Plates / 298
Column Base Plates / 300
Bearing on Milled Surfaces / 301
Plate Girders in Buildings / 302
Load Distribution to Bents and Shear Walls / 304
Combined Axial Compression or Tension and Bending / 306
Webs under Concentrated Loads / 308
Design of Stiffeners under Loads / 311
Fasteners for Buildings / 312
Composite Construction / 313
Number of Connectors Required for Building Construction / 316
Ponding Considerations in Buildings / 318

Chapter 10. Bridge and Suspension-Cable Formulas 321

Shear Strength Design for Bridges / 322
Allowable-Stress Design for Bridge Columns / 323
Load-and-Resistance Factor Design for Bridge Columns / 324
Allowable-Stress Design for Bridge Beams / 325
Stiffeners on Bridge Girders / 327
Hybrid Bridge Girders / 329

Load-Factor Design for Bridge Beams / *330*
Bearing on Milled Surfaces / *332*
Bridge Fasteners / *333*
Composite Construction in Highway Bridges / *333*
Number of Connectors in Bridges / *337*
Allowable-Stress Design for Shear in Bridges / *339*
Maximum Width/Thickness Ratios for Compression Elements for Highway Bridges / *341*
Suspension Cables / *341*
General Relations for Suspension Cables / *345*
Cable Systems / *353*

Chapter 11. Highway and Road Formulas 355

Circular Curves / *356*
Parabolic Curves / *359*
Highway Curves and Driver Safety / *361*
Highway Alignments / *362*
Structural Numbers for Flexible Pavements / *365*
Transition (Spiral) Curves / *370*
Designing Highway Culverts / *371*
American Iron and Steel Institute (AISI) Design Procedure / *374*

Chapter 12. Hydraulics and Waterworks Formulas 381

Capillary Action / *382*
Viscosity / *386*
Pressure on Submerged Curved Surfaces / *387*
Fundamentals of Fluid Flow / *388*
Similitude for Physical Models / *392*
Fluid Flow in Pipes / *395*
Pressure (Head) Changes Caused by Pipe Size Change / *403*
Flow through Orifices / *406*

Fluid Jets / *409*
Orifice Discharge into Diverging Conical Tubes / *410*
Water Hammer / *412*
Pipe Stresses Perpendicular to the Longitudinal Axis / *412*
Temperature Expansion of Pipe / *414*
Forces Due to Pipe Bends / *414*
Culverts / *417*
Open-Channel Flow / *420*
Manning's Equation for Open Channels / *424*
Hydraulic Jump / *425*
Nonuniform Flow in Open Channels / *429*
Weirs / *436*
Flow Over Weirs / *438*
Prediction of Sediment-Delivery Rate / *440*
Evaporation and Transpiration / *442*
Method for Determining Runoff for Minor Hydraulic Structures / *443*
Computing Rainfall Intensity / *443*
Groundwater / *446*
Water Flow for Firefighting / *446*
Flow from Wells / *447*
Economical Sizing of Distribution Piping / *448*
Venturi Meter Flow Computation / *448*
Hydroelectric Power Generation / *449*

Index 451

PREFACE

This handy book presents more than 2000 needed formulas for civil engineers to help them in the design office, in the field, and on a variety of construction jobs, anywhere in the world. These formulas are also useful to design drafters, structural engineers, bridge engineers, foundation builders, field engineers, professional-engineer license examination candidates, concrete specialists, timber-structure builders, and students in a variety of civil engineering pursuits.

The book presents formulas needed in 12 different specialized branches of civil engineering—beams and girders, columns, piles and piling, concrete structures, timber engineering, surveying, soils and earthwork, building structures, bridges, suspension cables, highways and roads, and hydraulics and open-channel flow. Key formulas are presented for each of these topics. Each formula is explained so the engineer, drafter, or designer knows how, where, and when to use the formula in professional work. Formula units are given in both the United States Customary System (USCS) and System International (SI). Hence, the text is usable throughout the world. To assist the civil engineer using this material in worldwide engineering practice, a comprehensive tabulation of conversion factors is presented in Chapter 1.

In assembling this collection of formulas, the author was guided by experts who recommended the areas of greatest need for a handy book of practical and applied civil engineering formulas.

Sources for the formulas presented here include the various regulatory and industry groups in the field of civil engineering, authors of recognized books on important topics in the field, drafters, researchers in the field of civil engineering, and a number of design engineers who work daily in the field of civil engineering. These sources are cited in the Acknowledgments.

When using any of the formulas in this book that may come from an industry or regulatory code, the user is cautioned to consult the latest version of the code. Formulas may be changed from one edition of a code to the next. In a work of this magnitude it is difficult to include the latest formulas from the numerous constantly changing codes. Hence, the formulas given here are those current at the time of publication of this book.

In a work this large it is possible that errors may occur. Hence, the author will be grateful to any user of the book who detects an error and calls it to the author's attention. Just write the author in care of the publisher. The error will be corrected in the next printing.

In addition, if a user believes that one or more important formulas have been left out, the author will be happy to consider them for inclusion in the next edition of the book. Again, just write him in care of the publisher.

Tyler G. Hicks, P.E.

ACKNOWLEDGMENTS

Many engineers, professional societies, industry associations, and governmental agencies helped the author find and assemble the thousands of formulas presented in this book. Hence, the author wishes to acknowledge this help and assistance.

The author's principal helper, advisor, and contributor was the late Frederick S. Merritt, P.E., Consulting Engineer. For many years Fred and the author were editors on companion magazines at The McGraw-Hill Companies. Fred was an editor on *Engineering-News Record,* whereas the author was an editor on *Power* magazine. Both lived on Long Island and traveled on the same railroad to and from New York City, spending many hours together discussing engineering, publishing, and book authorship.

When the author was approached by the publisher to prepare this book, he turned to Fred Merritt for advice and help. Fred delivered, preparing many of the formulas in this book and giving the author access to many more in Fred's extensive files and published materials. The author is most grateful to Fred for his extensive help, advice, and guidance.

Further, the author thanks the many engineering societies, industry associations, and governmental agencies whose work is referred to in this publication. These organizations provide the framework for safe design of numerous structures of many different types.

The author also thanks Larry Hager, Senior Editor, Professional Group, The McGraw-Hill Companies, for his excellent guidance and patience during the long preparation of the manuscript for this book. Finally, the author thanks his wife, Mary Shanley Hicks, a publishing professional, who always most willingly offered help and advice when needed.

Specific publications consulted during the preparation of this text include: American Association of State Highway and Transportation Officials (AASHTO) "Standard Specifications for Highway Bridges"; American Concrete Institute (ACI) "Building Code Requirements for Reinforced Concrete"; American Institute of Steel Construction (AISC) "Manual of Steel Construction," "Code of Standard Practice," and "Load and Resistance Factor Design Specifications for Structural Steel Buildings"; American Railway Engineering Association (AREA) "Manual for Railway Engineering"; American Society of Civil Engineers (ASCE) "Ground Water Management"; American Water Works Association (AWWA) "Water Quality and Treatment." In addition, the author consulted several hundred civil engineering reference and textbooks dealing with the topics in the current book. The author is grateful to the writers of all the publications cited here for the insight they gave him to civil engineering formulas. A number of these works are also cited in the text of this book.

HOW TO USE THIS BOOK

The formulas presented in this book are intended for use by civil engineers in every aspect of their professional work—design, evaluation, construction, repair, etc.

To find a suitable formula for the situation you face, start by consulting the index. Every effort has been made to present a comprehensive listing of all formulas in the book.

Once you find the formula you seek, read any accompanying text giving background information about the formula. Then when you understand the formula and its applications, insert the numerical values for the variables in the formula. Solve the formula and use the results for the task at hand.

Where a formula may come from a regulatory code, or where a code exists for the particular work being done, be certain to check the latest edition of the applicable code to see that the given formula agrees with the code formula. If it does not agree, be certain to use the latest code formula available. Remember, as a design engineer you are responsible for the structures you plan, design, and build. Using the latest edition of any governing code is the only sensible way to produce a safe and dependable design that you will be proud to be associated with. Further, you will sleep more peacefully!

CHAPTER 1

CONVERSION FACTORS FOR CIVIL ENGINEERING PRACTICE

Civil engineers throughout the world accept both the *United States Customary System* (USCS) and the *System International* (SI) units of measure for both applied and theoretical calculations. However, the SI units are much more widely used than those of the USCS. Hence, both the USCS and the SI units are presented for essentially every formula in this book. Thus, the user of the book can apply the formulas with confidence anywhere in the world.

To permit even wider use of this text, this chapter contains the conversion factors needed to switch from one system to the other. For engineers unfamiliar with either system of units, the author suggests the following steps for becoming acquainted with the unknown system:

1. *Prepare a list of measurements* commonly used in your daily work.
2. *Insert, opposite each known unit,* the unit from the other system. Table 1.1 shows such a list of USCS units with corresponding SI units and symbols prepared by a civil engineer who normally uses the USCS. The SI units shown in Table 1.1 were obtained from Table 1.3 by the engineer.
3. *Find, from a table of conversion factors,* such as Table 1.3, the value used to convert from USCS to SI units. Insert each appropriate value in Table 1.2 from Table 1.3.
4. *Apply the conversion values* wherever necessary for the formulas in this book.
5. *Recognize—here and now*—that the most difficult aspect of becoming familiar with a new system of measurement is becoming comfortable with the names and magnitudes of the units. Numerical conversion is simple, once you have set up your own conversion table.

TABLE 1.1 Commonly Used USCS and SI Units[†]

USCS unit	SI unit	SI symbol	Conversion factor (multiply USCS unit by this factor to obtain SI unit)
square foot	square meter	m^2	0.0929
cubic foot	cubic meter	m^3	0.2831
pound per square inch	kilopascal	kPa	6.894
pound force	newton	Nu	4.448
foot pound torque	newton meter	N·m	1.356
kip foot	kilonewton meter	kN·m	1.355
gallon per minute	liter per second	L/s	0.06309
kip per square inch	megapascal	MPa	6.89

[†]This table is abbreviated. For a typical engineering practice, an actual table would be many times this length.

Be careful, when using formulas containing a numerical constant, to convert the constant to that for the system you are using. You can, however, use the formula for the USCS units (when the formula is given in those units) and then convert the final result to the SI equivalent using Table 1.3. For the few formulas given in SI units, the reverse procedure should be used.

TABLE 1.2 Typical Conversion Table[†]

To convert from	To	Multiply by[‡]	
square foot	square meter	9.290304	E − 02
foot per second squared	meter per second squared	3.048	E − 01
cubic foot	cubic meter	2.831685	E − 02
pound per cubic inch	kilogram per cubic meter	2.767990	E + 04
gallon per minute	liter per second	6.309	E − 02
pound per square inch	kilopascal	6.894757	
pound force	newton	4.448222	
kip per square foot	pascal	4.788026	E + 04
acre foot per day	cubic meter per second	1.427641	E − 02
acre	square meter	4.046873	E + 03
cubic foot per second	cubic meter per second	2.831685	E − 02

[†]This table contains only selected values. See the U.S. Department of the Interior *Metric Manual*, or National Bureau of Standards, *The International System of Units* (SI), both available from the U.S. Government Printing Office (GPO), for far more comprehensive listings of conversion factors.
[‡]The E indicates an exponent, as in scientific notation, followed by a positive or negative number, representing the power of 10 by which the given conversion factor is to be multiplied before use. Thus, for the square foot conversion factor, 9.290304 × 1/100 = 0.09290304, the factor to be used to convert square feet to square meters. For a positive exponent, as in converting acres to square meters, multiply by 4.046873 × 1000 = 4046.8.

Where a conversion factor cannot be found, simply use the dimensional substitution. Thus, to convert pounds per cubic inch to kilograms per cubic meter, find 1 lb = 0.4535924 kg and 1 in^3 = 0.00001638706 m^3. Then, 1 lb/in^3 = 0.4535924 kg/0.00001638706 m^3 = 27,680.01, or 2.768 E + 4.

TABLE 1.3 Factors for Conversion to SI Units of Measurement

To convert from	To	Multiply by
acre foot, acre ft	cubic meter, m^3	1.233489 E + 03
acre	square meter, m^2	4.046873 E + 03
angstrom, Å	meter, m	1.000000* E − 10
atmosphere, atm (standard)	pascal, Pa	1.013250* E + 05
atmosphere, atm (technical = 1 kgf/cm^2)	pascal, Pa	9.806650* E + 04
bar	pascal, Pa	1.000000* E + 05
barrel (for petroleum, 42 gal)	cubic meter, m^2	1.589873 E − 01
board foot, board ft	cubic meter, m^3	2.359737 E − 03
British thermal unit, Btu, (mean)	joule, J	1.05587 E + 03
British thermal unit, Btu (International Table)·in/(h)(ft^2)(°F) (*k*, thermal conductivity)	watt per meter kelvin, W/(m·K)	1.442279 E − 01
British thermal unit, Btu (International Table)/h	watt, W	2.930711 E − 01
British thermal unit, Btu (International Table)/(h)(ft^2)(°F) (*C*, thermal conductance)	watt per square meter kelvin, W/(m^2·K)	5.678263 E + 00
British thermal unit, Btu (International Table)/lb	joule per kilogram, J/kg	2.326000* E + 03

TABLE 1.3 Factors for Conversion to SI Units of Measurement *(Continued)*

To convert from	To	Multiply by
British thermal unit, Btu (International Table)/(lb)(°F) (*c*, heat capacity)	joule per kilogram kelvin, J/(kg·K)	4.186800* E + 03
British thermal unit, cubic foot, Btu (International Table)/ft^3	joule per cubic meter, J/m^3	3.725895 E + 04
bushel (U.S.)	cubic meter, m^3	3.523907 E − 02
calorie (mean)	joule, J	4.19002 E + 00
candela per square inch, cd/in^2	candela per square meter, cd/m^2	1.550003 E + 03
centimeter, cm, of mercury (0°C)	pascal, Pa	1.33322 E + 03
centimeter, cm, of water (4°C)	pascal, Pa	9.80638 E + 01
chain	meter, m	2.011684 E + 01
circular mil	square meter, m^2	5.067075 E − 10
day	second, s	8.640000* E + 04
day (sidereal)	second, s	8.616409 E + 04
degree (angle)	radian, rad	1.745329 E − 02
degree Celsius	kelvin, K	$T_K = t_C + 273.15$
degree Fahrenheit	degree Celsius, °C	$t_C = (t_F - 32)/1.8$
degree Fahrenheit	kelvin, K	$T_K = (t_F + 459.67)/1.8$
degree Rankine	kelvin, K	$T_K = T_R/1.8$
(°F)(h)(ft^2)/Btu (International Table) (*R*, thermal resistance)	kelvin square meter per watt, K·m^2/W	1.761102 E − 01

TABLE 1.3 Factors for Conversion to SI Units of Measurement *(Continued)*

To convert from	To	Multiply by
(°F)(h)(ft^2)/(Btu (International Table)·in) (thermal resistivity)	kelvin meter per watt, K·m/W	6.933471 E + 00
dyne, dyn	newton, N	1.000000† E − 05
fathom	meter, m	1.828804 E + 00
foot, ft	meter, m	3.048000† E − 01
foot, ft (U.S. survey)	meter, m	3.048006 E − 01
foot, ft, of water (39.2°F) (pressure)	pascal, Pa	2.98898 E + 03
square foot, ft^2	square meter, m^2	9.290304† E − 02
square foot per hour, ft^2/h (thermal diffusivity)	square meter per second, m^2/s	2.580640† E − 05
square foot per second, ft^2/s	square meter per second, m^2/s	9.290304† E − 02
cubic foot, ft^3 (volume or section modulus)	cubic meter, m^3	2.831685 E − 02
cubic foot per minute, ft^3/min	cubic meter per second, m^3/s	4.719474 E − 04
cubic foot per second, ft^3/s	cubic meter per second, m^3/s	2.831685 E − 02
foot to the fourth power, ft^4 (area moment of inertia)	meter to the fourth power, m^4	8.630975 E − 03
foot per minute, ft/min	meter per second, m/s	5.080000† E − 03
foot per second, ft/s	meter per second, m/s	3.048000† E − 01

TABLE 1.3 Factors for Conversion to SI Units of Measurement *(Continued)*

To convert from	To	Multiply by
foot per second squared, ft/s^2	meter per second squared, m/s^2	3.048000† E − 01
footcandle, fc	lux, lx	1.076391 E + 01
footlambert, fL	candela per square meter, cd/m^2	3.426259 E + 00
foot pound force, ft·lbf	joule, J	1.355818 E + 00
foot pound force per minute, ft·lbf/min	watt, W	2.259697 E − 02
foot pound force per second, ft·lbf/s	watt, W	1.355818 E + 00
foot poundal, ft poundal	joule, J	4.214011 E − 02
free fall, standard *g*	meter per second squared, m/s^2	9.806650† E + 00
gallon, gal (Canadian liquid)	cubic meter, m^3	4.546090 E − 03
gallon, gal (U.K. liquid)	cubic meter, m^3	4.546092 E − 03
gallon, gal (U.S. dry)	cubic meter, m^3	4.404884 E − 03
gallon, gal (U.S. liquid)	cubic meter, m^3	3.785412 E − 03
gallon, gal (U.S. liquid) per day	cubic meter per second, m^3/s	4.381264 E − 08
gallon, gal (U.S. liquid) per minute	cubic meter per second, m^3/s	6.309020 E − 05
grad	degree (angular)	9.000000† E − 01
grad	radian, rad	1.570796 E − 02
grain, gr	kilogram, kg	6.479891† E − 05
gram, g	kilogram, kg	1.000000† E − 03

TABLE 1.3 Factors for Conversion to SI Units of Measurement *(Continued)*

To convert from	To	Multiply by
hectare, ha	square meter, m^2	1.000000† E + 04
horsepower, hp (550 ft·lbf/s)	watt, W	7.456999 E + 02
horsepower, hp (boiler)	watt, W	9.80950 E + 03
horsepower, hp (electric)	watt, W	7.460000† E + 02
horsepower, hp (water)	watt, W	7.46043† E + 02
horsepower, hp (U.K.)	watt, W	7.4570 E + 02
hour, h	second, s	3.600000† E + 03
hour, h (sidereal)	second, s	3.590170 E + 03
inch, in	meter, m	2.540000† E − 02
inch of mercury, in Hg (32°F) (pressure)	pascal, Pa	3.38638 E + 03
inch of mercury, in Hg (60°F) (pressure)	pascal, Pa	3.37685 E + 03
inch of water, in H_2O (60°F) (pressure)	pascal, Pa	2.4884 E + 02
square inch, in^2	square meter, m^2	6.451600† E − 04
cubic inch, in^3 (volume or section modulus)	cubic meter, m^3	1.638706 E − 05
inch to the fourth power, in^4 (area moment of inertia)	meter to the fourth power, m^4	4.162314 E − 07
inch per second, in/s	meter per second, m/s	2.540000† E − 02

TABLE 1.3 Factors for Conversion to SI Units of Measurement *(Continued)*

To convert from	To	Multiply by
kelvin, K	degree Celsius, °C	$t_C = T_K - 273.15$
kilogram force, kgf	newton, N	9.806650† E + 00
kilogram force meter, kg·m	newton meter, N·m	9.806650† E + 00
kilogram force second squared per meter, kgf·s^2/m (mass)	kilogram, kg	9.806650† E + 00
kilogram force per square centimeter, kgf/cm^2	pascal, Pa	9.806650† E + 04
kilogram force per square meter, kgf/m^2	pascal, Pa	9.806650† E + 00
kilogram force per square millimeter, kgf/mm^2	pascal, Pa	9.806650† E + 06
kilometer per hour, km/h	meter per second, m/s	2.777778 E − 01
kilowatt hour, kWh	joule, J	3.600000† E + 06
kip (1000 lbf)	newton, N	4.448222 E + 03
kipper square inch, kip/in^2 ksi	pascal, Pa	6.894757 E + 06
knot, kn (international)	meter per second, m/s	5.144444 E − 01
lambert, L	candela per square meter, cd/m	3.183099 E + 03
liter	cubic meter, m^3	1.000000† E − 03
maxwell	weber, Wb	1.000000† E − 08
mho	siemens, S	1.000000† E + 00

TABLE 1.3 Factors for Conversion to SI Units of Measurement *(Continued)*

To convert from	To	Multiply by
microinch, μin	meter, m	2.540000† E − 08
micron, μm	meter, m	1.000000† E − 06
mil, mi	meter, m	2.540000† E − 05
mile, mi (international)	meter, m	1.609344† E + 03
mile, mi (U.S. statute)	meter, m	1.609347 E + 03
mile, mi (international nautical)	meter, m	1.852000† E + 03
mile, mi (U.S. nautical)	meter, m	1.852000† E + 03
square mile, mi^2 (international)	square meter, m^2	2.589988 E + 06
square mile, mi^2 (U.S. statute)	square meter, m^2	2.589998 E + 06
mile per hour, mi/h (international)	meter per second, m/s	4.470400† E − 01
mile per hour, mi/h (international)	kilometer per hour, km/h	1.609344† E + 00
millibar, mbar	pascal, Pa	1.000000† E + 02
millimeter of mercury, mmHg (0°C)	pascal, Pa	1.33322 E + 02
minute (angle)	radian, rad	2.908882 E − 04
minute, min	second, s	6.000000† E + 01
minute (sidereal)	second, s	5.983617 E + 01
ounce, oz (avoirdupois)	kilogram, kg	2.834952 E − 02
ounce, oz (troy or apothecary)	kilogram, kg	3.110348 E − 02
ounce, oz (U.K. fluid)	cubic meter, m^3	2.841307 E − 05
ounce, oz (U.S. fluid)	cubic meter, m^3	2.957353 E − 05
ounce force, ozf	newton, N	2.780139 E − 01

TABLE 1.3 Factors for Conversion to SI Units of Measurement *(Continued)*

To convert from	To	Multiply by
ounce force·inch, ozf·in	newton meter, N·m	7.061552 E − 03
ounce per square foot, oz (avoirdupois)/ft^2	kilogram per square meter, kg/m^2	3.051517 E − 01
ounce per square yard, oz (avoirdupois)/yd^2	kilogram per square meter, kg/m^2	3.390575 E − 02
perm (0°C)	kilogram per pascal second meter, kg/(Pa·s·m)	5.72135 E − 11
perm (23°C)	kilogram per pascal second meter, kg/(Pa·s·m)	5.74525 E − 11
perm inch, perm·in (0°C)	kilogram per pascal second meter, kg/(Pa·s·m)	1.45322 E − 12
perm inch, perm·in (23°C)	kilogram per pascal second meter, kg/(Pa·s·m)	1.45929 E − 12
pint, pt (U.S. dry)	cubic meter, m^3	5.506105 E − 04
pint, pt (U.S. liquid)	cubic meter, m^3	4.731765 E − 04
poise, p (absolute viscosity)	pascal second, Pa·s	1.000000† E − 01
pound, lb (avoirdupois)	kilogram, kg	4.535924 E − 01
pound, lb (troy or apothecary)	kilogram, kg	3.732417 E − 01
pound square inch, lb·in^2 (moment of inertia)	kilogram square meter, kg·m^2	2.926397 E − 04

TABLE 1.3 Factors for Conversion to SI Units of Measurement *(Continued)*

To convert from	To	Multiply by
pound per foot·second, lb/ft·s	pascal second, Pa·s	1.488164 E + 00
pound per square foot, lb/ft^2	kilogram per square meter, kg/m^2	4.882428 E + 00
pound per cubic foot, lb/ft^3	kilogram per cubic meter, kg/m^3	1.601846 E − 01
pound per gallon, lb/gal (U.K. liquid)	kilogram per cubic meter, kg/m^3	9.977633 E + 01
pound per gallon, lb/gal (U.S. liquid)	kilogram per cubic meter, kg/m^3	1.198264 E + 02
pound per hour, lb/h	kilogram per second, kg/s	1.259979 E − 04
pound per cubic inch, lb/in^3	kilogram per cubic meter, kg/m^3	2.767990 E + 04
pound per minute, lb/min	kilogram per second, kg/s	7.559873 E − 03
pound per second, lb/s	kilogram per second, kg/s	4.535924 E − 01
pound per cubic yard, lb/yd^3	kilogram per cubic meter, kg/m^3	5.932764 E − 01
poundal	newton, N	1.382550 E − 01
pound·force, lbf	newton, N	4.448222 E + 00
pound force foot, lbf·ft	newton meter, N·m	1.355818 E + 00
pound force per foot, lbf/ft	newton per meter, N/m	1.459390 E + 01
pound force per square foot, lbf/ft^2	pascal, Pa	4.788026 E + 01
pound force per inch, lbf/in	newton per meter, N/m	1.751268 E + 02

TABLE 1.3 Factors for Conversion to SI Units of Measurement *(Continued)*

To convert from	To	Multiply by
pound force per square inch, lbf/in^2 (psi)	pascal, Pa	6.894757 E + 03
quart, qt (U.S. dry)	cubic meter, m^3	1.101221 E − 03
quart, qt (U.S. liquid)	cubic meter, m^3	9.463529 E − 04
rod	meter, m	5.029210 E + 00
second (angle)	radian, rad	4.848137 E − 06
second (sidereal)	second, s	9.972696 E − 01
square (100 ft^2)	square meter, m^2	9.290304† E + 00
ton (assay)	kilogram, kg	2.916667 E − 02
ton (long, 2240 lb)	kilogram, kg	1.016047 E + 03
ton (metric)	kilogram, kg	1.000000† E + 03
ton (refrigeration)	watt, W	3.516800 E + 03
ton (register)	cubic meter, m^3	2.831685 E + 00
ton (short, 2000 lb)	kilogram, kg	9.071847 E + 02
ton (long per cubic yard, ton)/yd^3	kilogram per cubic meter, kg/m^3	1.328939 E + 03
ton (short per cubic yard, ton)/yd^3	kilogram per cubic meter, kg/m^3	1.186553 E + 03
ton force (2000 lbf)	newton, N	8.896444 E + 03
tonne, t	kilogram, kg	1.000000† E + 03
watt hour, Wh	joule, J	3.600000† E + 03
yard, yd	meter, m	9.144000† E − 01
square yard, yd^2	square meter, m^2	8.361274 E − 01
cubic yard, yd^3	cubic meter, m^3	7.645549 E − 01
year (365 days)	second, s	3.153600† E + 07
year (sidereal)	second, s	3.155815 E + 07

†Exact value.
From E380, "Standard for Metric Practice," American Society for Testing and Materials.

In analyzing beams of various types, the geometric properties of a variety of cross-sectional areas are used. Figure 2.1 gives equations for computing area A, moment of inertia I, section modulus or the ratio $S = I/c$, where c = distance from the neutral axis to the outermost fiber of the beam or other member. Units used are inches and millimeters and their powers. The formulas in Fig. 2.1 are valid for both USCS and SI units.

Handy formulas for some dozen different types of beams are given in Fig. 2.2. In Fig. 2.2, both USCS and SI units can be used in any of the formulas that are applicable to both steel and wooden beams. Note that W = load, lb (kN); L = length, ft (m); R = reaction, lb (kN); V = shear, lb (kN); M = bending moment, lb · ft (N · m); D = deflection, ft (m); a = spacing, ft (m); b = spacing, ft (m); E = modulus of elasticity, lb/in^2 (kPa); I = moment of inertia, in^4 (dm^4); $<$ = less than; $>$ = greater than.

Figure 2.3 gives the elastic-curve equations for a variety of prismatic beams. In these equations the load is given as P, lb (kN). Spacing is given as k, ft (m) and c, ft (m).

CONTINUOUS BEAMS

Continuous beams and frames are statically indeterminate. Bending moments in these beams are functions of the geometry, moments of inertia, loads, spans, and modulus of elasticity of individual members. Figure 2.4 shows how any span of a continuous beam can be treated as a single beam, with the moment diagram decomposed into basic components. Formulas for analysis are given in the diagram. Reactions of a continuous beam can be found by using the formulas in Fig. 2.5. Fixed-end moment formulas for beams of constant moment of inertia (prismatic beams) for

CHAPTER 2

BEAM FORMULAS

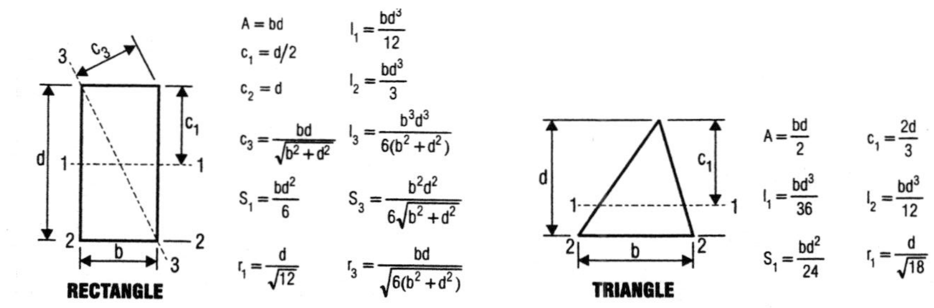

RECTANGLE
$A = bd$
$c_1 = d/2$
$c_2 = d$
$c_3 = \frac{bd}{\sqrt{b^2 + d^2}}$
$S_1 = \frac{bd^2}{6}$
$r_1 = \frac{d}{\sqrt{12}}$
$I_1 = \frac{bd^3}{12}$
$I_2 = \frac{bd^3}{3}$
$I_3 = \frac{b^3d^3}{6(b^2 + d^2)}$
$S_3 = \frac{b^2d^2}{6\sqrt{b^2 + d^2}}$
$r_3 = \frac{bd}{\sqrt{6(b^2 + d^2)}}$

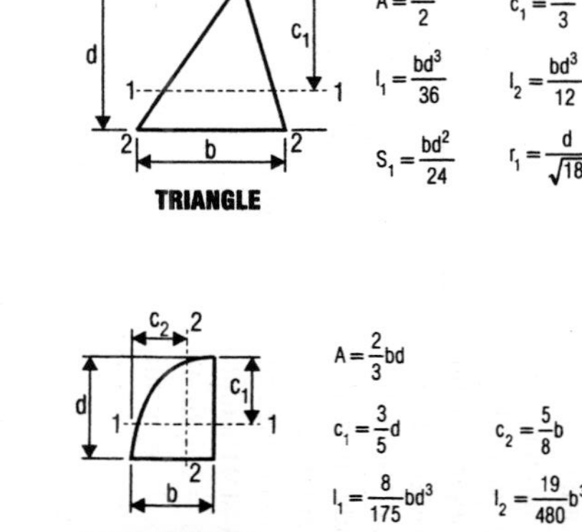

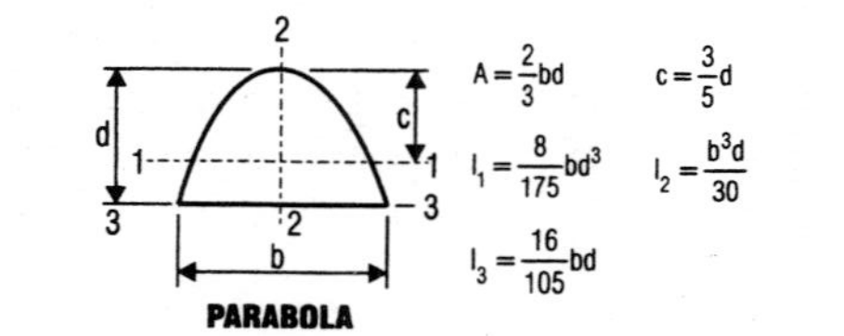

FIGURE 2.1 Geometric properties of sections.

Section	Moment of inertia	Section modulus	Radius of gyration
Equilateral Polygon A = area R = rad circumscribed circle r = rad inscribed circle n = no. sides a = length of side Axis as in preceding section of octagon	$I = \frac{A}{24}(6R^2 - a^2)$ $= \frac{A}{48}(12r^2 + a^2)$ $= \frac{AR^2}{4}$ (approx)	$\frac{I}{c} = \frac{I}{r}$ $= \frac{I}{R \cos \frac{180°}{n}}$ $= \frac{AR}{4}$ (approx)	$\sqrt{\frac{6R^2 - a^2}{24}} \approx \frac{R}{2}$ $\sqrt{\frac{12r^2 + a^2}{48}}$
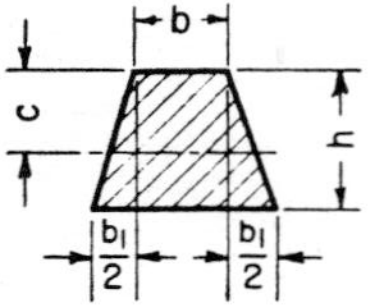	$I = \frac{6b^2 + 6bb_1 + b_1^2}{36(2b + b_1)} h^3$ $c = \frac{1}{3} \frac{3b + 2b_1}{2b + b_1} h$	$\frac{I}{c} = \frac{6b^2 + 6bb_1 + b_1^2}{12(3b + 2b_1)} h^2$	$\frac{h\sqrt{12b^2 + 12bb_1 + 2b_1^2}}{6(2b + b_1)}$

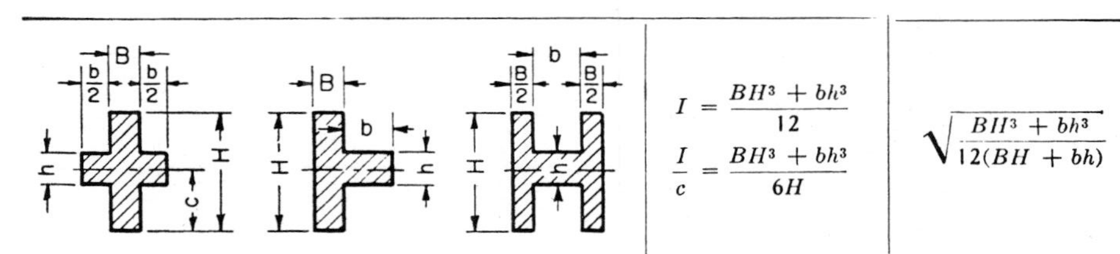

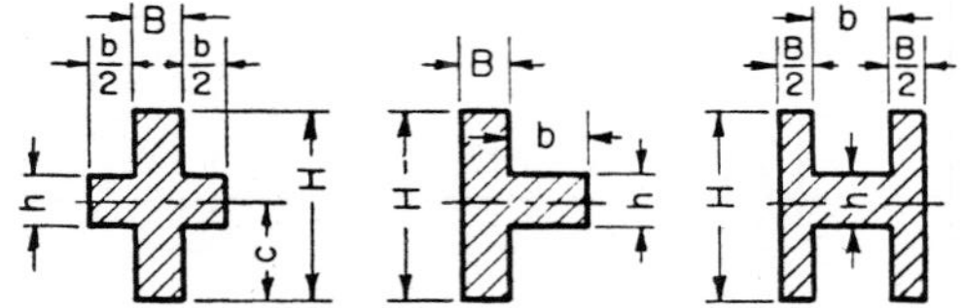	$I = \dfrac{BH^3 + bh^3}{12}$ $\dfrac{I}{c} = \dfrac{BH^3 + bh^3}{6H}$	$\sqrt{\dfrac{BH^3 + bh^3}{12(BH + bh)}}$
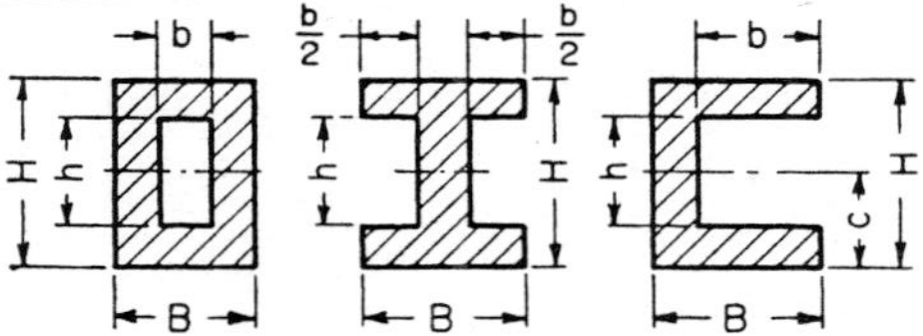	$I = \dfrac{BH^3 - bh^3}{12}$ $\dfrac{I}{c} = \dfrac{BH^3 - bh^3}{6H}$	$\sqrt{\dfrac{BH^3 - bh^3}{12(BH - bh)}}$

FIGURE 2.1 (*Continued*) Geometric properties of sections.

Section	Moment of inertia and section modulus	Radius of gyration
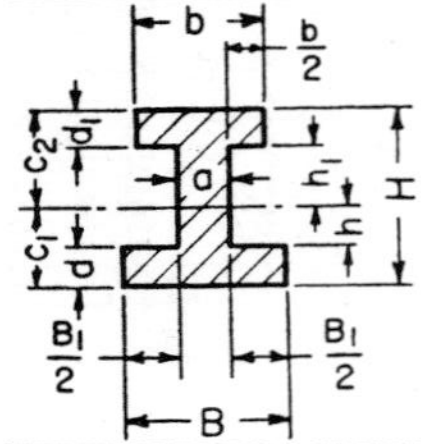	$I = \frac{1}{3}(Bc_1^3 - B_1h^3 + bc_2^3 - b_1h_1^3)$ $c_1 = \frac{1}{2}\frac{aH^2 + B_1d^2 + b_1d_1(2H - d_1)}{aH + B_1d + b_1d_1}$	$\sqrt{\frac{I}{(Bd + bd_1) + a(h + h_1)}}$
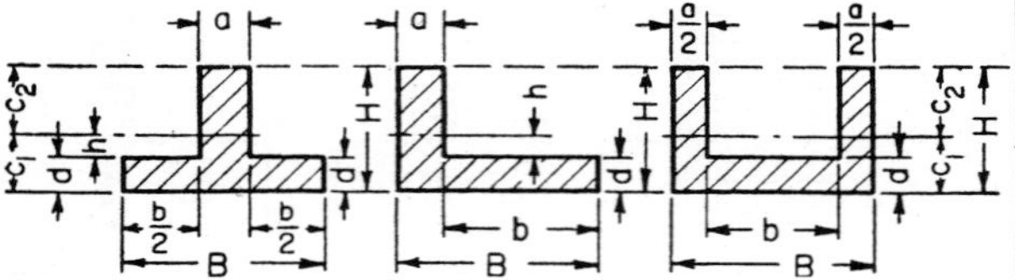		$I = \frac{1}{3}(Bc_1^3 - bh^3 + ac_2^3)$ $c_1 = \frac{1}{2}\frac{aH^2 + bd^2}{aH + bd}$ $c_2 = H - c_1$ $r = \sqrt{\frac{I}{[Bd + a(H - d)]}}$

Section	Moment of inertia	Section modulus	Radius of gyration
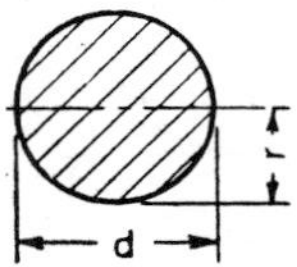	$I = \frac{\pi d^4}{64} = \frac{\pi r^4}{4} = \frac{A}{4} r^2$ $= 0.05d^4$ (approx)	$\frac{I}{c} = \frac{\pi d^3}{32} = \frac{\pi r^3}{4} = \frac{A}{4} r$ $= 0.1d^3$ (approx)	$\frac{r}{2} = \frac{d}{4}$
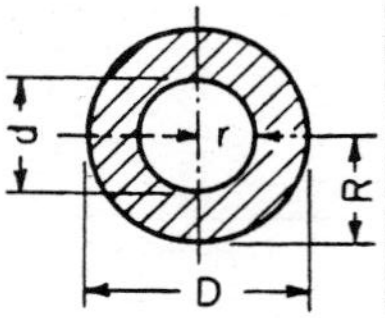 $d_m = \tfrac{1}{2}(D + d)$ $s = \tfrac{1}{2}(D - d)$	$I = \frac{\pi}{64}(D^4 - d^4)$ $= \frac{\pi}{4}(R^4 - r^4)$ $= \tfrac{1}{4}A(R^2 + r^2)$ $= 0.05(D^4 - d^4)$ (approx)	$\frac{I}{c} = \frac{\pi}{32}\frac{D^4 - d^4}{D}$ $= \frac{\pi}{4}\frac{R^4 - r^4}{R}$ $= 0.8d_m^2 s$ (approx) when $\frac{s}{d_m}$ is very small	$\frac{\sqrt{R^2 + r^2}}{2} = \frac{\sqrt{D^2 + d^2}}{4}$

FIGURE 2.1 (*Continued*) Geometric properties of sections.

Section	Moment of inertia	Section modulus	Radius of gyration
c_1, c_2, r	$I = r^4\left(\frac{\pi}{8} - \frac{8}{9\pi}\right)$ $= 0.1098r^4$	$\frac{I}{c_2} = 0.1908r^3$ $\frac{I}{c_1} = 0.2587r^3$ $c_1 = 0.4244r$	$\frac{\sqrt{9\pi^2 - 64}}{6\pi} r = 0.264r$
c_1, c_2, t, r, r_1, R	$I = 0.1098(R^4 - r^4) - \frac{0.283R^2r^2(R - r)}{R + r}$ $= 0.3tr_1^3$ (approx) when $\frac{t}{r_1}$ is very small	$c_1 = \frac{4}{3\pi} \frac{R^2 + Rr + r^2}{R + r}$ $c_2 = R - c_1$	$\sqrt{\frac{2I}{\pi(R^2 - r^2)}}$ $= 0.31r_1$ (approx)

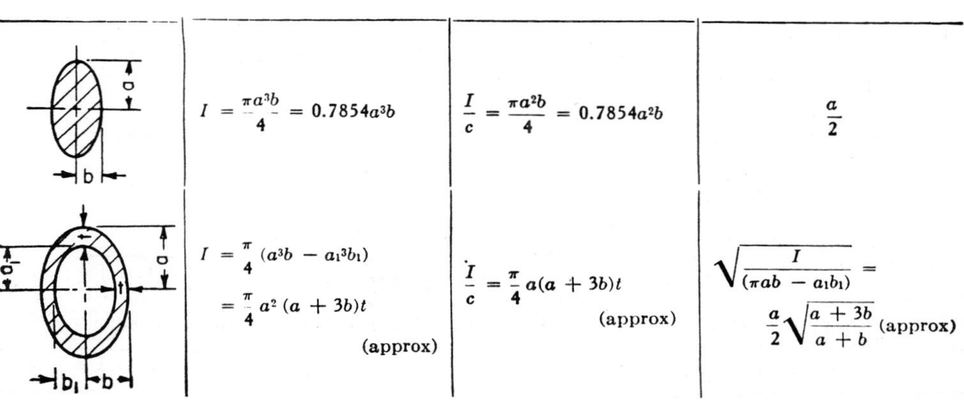

	$I = \frac{\pi a^3 b}{4} = 0.7854a^3b$	$\frac{I}{c} = \frac{\pi a^2 b}{4} = 0.7854a^2b$	$\frac{a}{2}$
	$I = \frac{\pi}{4}(a^3b - a_1^3b_1)$ $= \frac{\pi}{4}a^2(a + 3b)t$ (approx)	$\frac{I}{c} = \frac{\pi}{4}a(a + 3b)t$ (approx)	$\sqrt{\frac{I}{(\pi ab - a_1b_1)}} = \frac{a}{2}\sqrt{\frac{a + 3b}{a + b}}$ (approx)

FIGURE 2.1 *(Continued)* Geometric properties of sections.

Section	Moment of inertia and section modulus	Radius of gyration
b, h, d, b	$I = \frac{1}{12}\left[\frac{3\pi}{16}d^4 + b(h^3 - d^3) + b^3(h - d)\right]$ $\frac{I}{c} = \frac{1}{6h}\left[\frac{3\pi}{16}d^4 + b(h^3 + d^3) + b^3(h - d)\right]$	$\sqrt{\frac{I}{\pi\frac{d^2}{4} + 2b(h - d)}}$ (approx)
$\frac{B}{2}$, $\frac{B}{2}$, $\frac{B}{4}$, h, $\frac{B}{4}$, H, t, B	$I = \frac{t}{4}\left(\frac{\pi B^3}{16} + B^2h + \frac{\pi Bh^2}{2} + \frac{2}{3}h^3\right)$ $h = H - \frac{1}{2}B$ $\frac{I}{c} = \frac{2I}{H + t}$	$\sqrt{\frac{I}{2\left(\frac{\pi B}{4} + h\right)t}}$

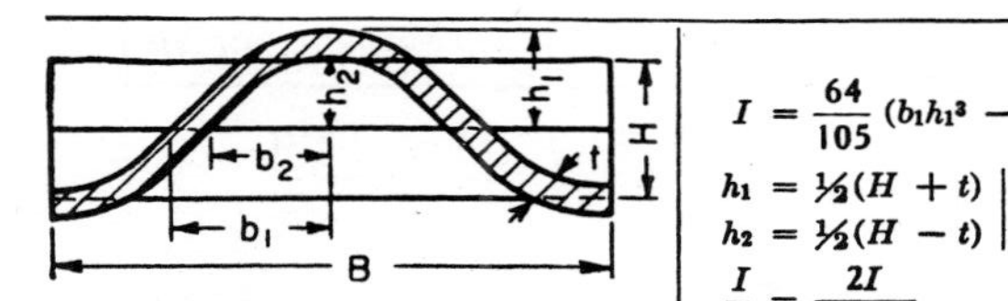

Corrugated sheet iron, parabolically curved

$$I = \frac{64}{105}(b_1 h_1^3 - b_2 h_2^3), \text{ where}$$

$$h_1 = \tfrac{1}{2}(H + t) \quad b_1 = \tfrac{1}{4}(B + 2.6t)$$

$$h_2 = \tfrac{1}{2}(H - t) \quad b_2 = \tfrac{1}{4}(B - 2.6t)$$

$$\frac{I}{c} = \frac{2I}{H + t}$$

$$r = \sqrt{\frac{3I}{t(2B + 5.2H)}}$$

FIGURE 2.1 (*Continued*) Geometric properties of sections.

Approximate values of *least* radius of gyration r

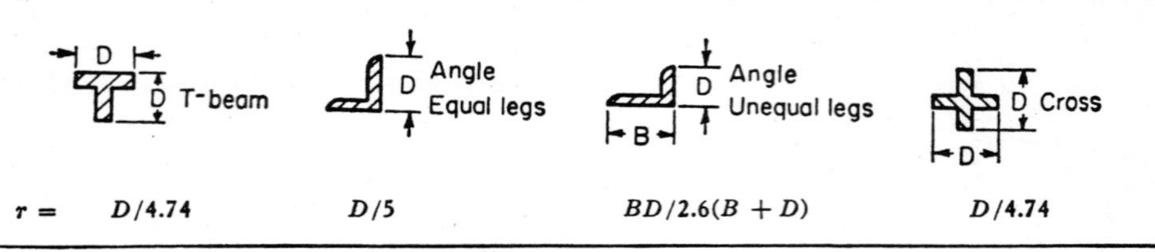

	Phoenix column	Carnegie Z-bar column	I-beam	Channel	Deck beam
$r =$	$0.3636D$	$0.295D$	$D/4.58$	$D/3.54$	$D/6$

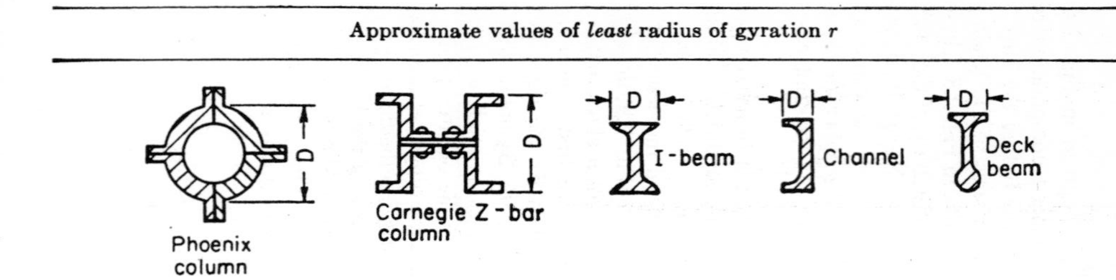

	T-beam	Angle, Equal legs	Angle, Unequal legs	Cross
$r =$	$D/4.74$	$D/5$	$BD/2.6(B + D)$	$D/4.74$

FIGURE 2.1 (*Continued*) Geometric properties of sections.

several common types of loading are given in Fig. 2.6. Curves (Fig. 2.7) can be used to speed computation of fixed-end moments in prismatic beams. Before the curves in Fig. 2.7 can be used, the characteristics of the loading must be computed by using the formulas in Fig. 2.8. These include $\bar{x}L$, the location of the center of gravity of the loading with respect to one of the loads; $G^2 = \Sigma b_n^2 P_n/W$, where b_nL is the distance from each load P_n to the center of gravity of the loading (taken positive to the right); and $S^3 = \Sigma b_n^3 P_n/W$. These values are given in Fig. 2.8 for some common types of loading.

Formulas for moments due to deflection of a fixed-end beam are given in Fig. 2.9. To use the modified moment distribution method for a fixed-end beam such as that in Fig. 2.9, we must first know the fixed-end moments for a beam with supports at different levels. In Fig. 2.9, the right end of a beam with span L is at a height d above the left end. To find the fixed-end moments, we first deflect the beam with both ends hinged; and then fix the right end, leaving the left end hinged, as in Fig. 2.9*b*. By noting that a line connecting the two supports makes an angle approximately equal to d/L (its tangent) with the original position of the beam, we apply a moment at the hinged end to produce an end rotation there equal to d/L. By the definition of stiffness, this moment equals that shown at the left end of Fig. 2.9*b*. The carryover to the right end is shown as the top formula on the right-hand side of Fig. 2.9*b*. By using the law of reciprocal deflections, we obtain the end moments of the deflected beam in Fig. 2.9 as

$$M_L^F = K_L^F(1 + C_R^F)\frac{d}{L} \tag{2.1}$$

$$M_R^F = K_R^F(1 + C_L^F)\frac{d}{L} \tag{2.2}$$

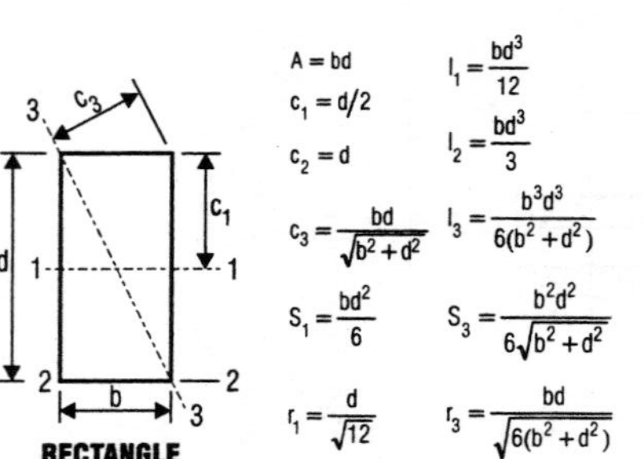

RECTANGLE

$$A = bd \qquad I_1 = \frac{bd^3}{12}$$

$$c_1 = d/2 \qquad I_2 = \frac{bd^3}{3}$$

$$c_2 = d$$

$$c_3 = \frac{bd}{\sqrt{b^2 + d^2}} \qquad I_3 = \frac{b^3 d^3}{6(b^2 + d^2)}$$

$$S_1 = \frac{bd^2}{6} \qquad S_3 = \frac{b^2 d^2}{6\sqrt{b^2 + d^2}}$$

$$r_1 = \frac{d}{\sqrt{12}} \qquad r_3 = \frac{bd}{\sqrt{6(b^2 + d^2)}}$$

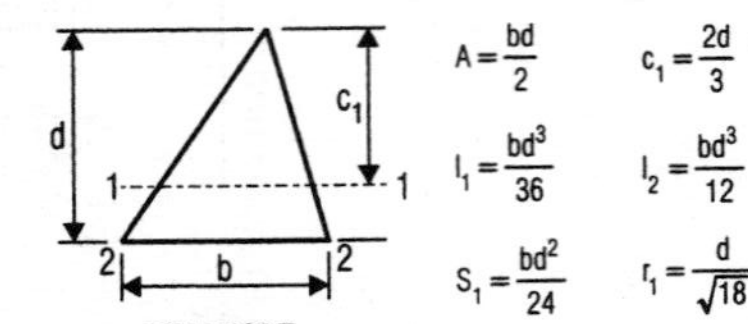

TRIANGLE

$$A = \frac{bd}{2} \qquad c_1 = \frac{2d}{3}$$

$$I_1 = \frac{bd^3}{36} \qquad I_2 = \frac{bd^3}{12}$$

$$S_1 = \frac{bd^2}{24} \qquad r_1 = \frac{d}{\sqrt{18}}$$

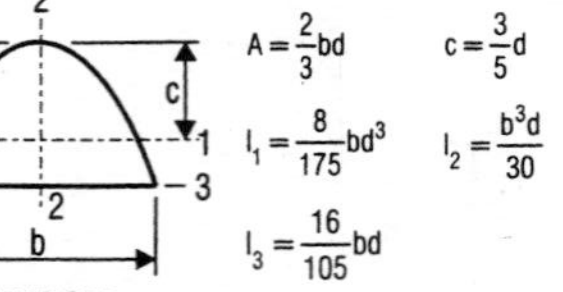

PARABOLA

$$A = \frac{2}{3}bd \qquad c = \frac{3}{5}d$$

$$I_1 = \frac{8}{175}bd^3 \qquad I_2 = \frac{b^3 d}{30}$$

$$I_3 = \frac{16}{105}bd$$

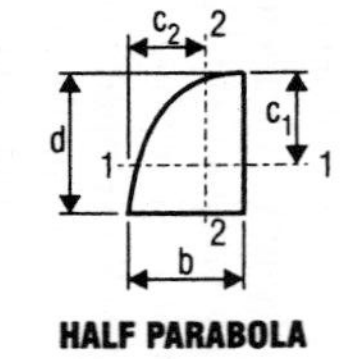
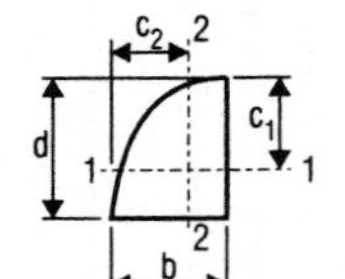

HALF PARABOLA

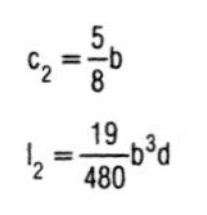
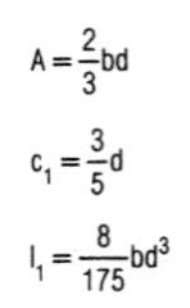

$$A = \frac{2}{3}bd$$

$$c_1 = \frac{3}{5}d \qquad c_2 = \frac{5}{8}b$$

$$I_1 = \frac{8}{175}bd^3 \qquad I_2 = \frac{19}{480}b^3 d$$

CASE 2. - Beam Supported Both Ends - Concentrated Load at Any Point.

$R = \frac{Wb}{L}$ $R_1 = \frac{Wa}{L}$ $V(\text{max.}) = R$ when $a<b$ and R_1 when $a>b$ At x: $V = \frac{Wb}{L}$	At point of load: $M(\text{max.}) = \frac{Wab}{L}$ At x: when $x<a$ $M = \frac{Wbx}{L}$	At x: when $x = \sqrt{a(a+2b) \div 3}$ and $a>b$ $D(\text{max.}) = Wab(a+2b)\sqrt{3a(a+2b)} \div 27\ EIL$ At x: when $x<a$ $D = \frac{Wbx}{6\ EIL}\left[2L(L-x) - b^2 - (L-x)^2\right]$ At x: when $x>a$ $D = \frac{Wa(L-x)}{6\ EIL}\left[2Lb - b^2 - (L-x)^2\right]$

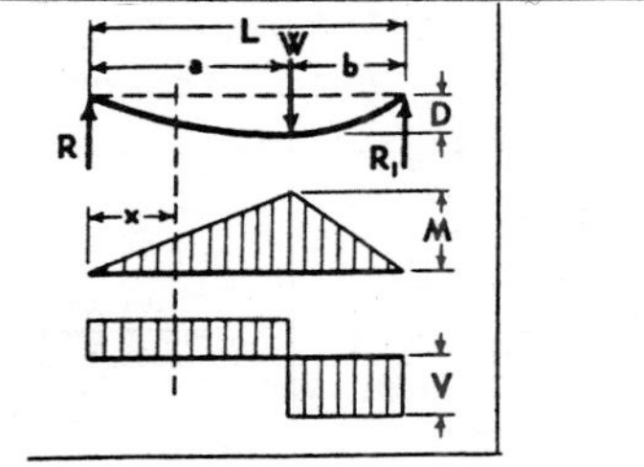

FIGURE 2.2 Beam formulas. *(From J. Callender, Time-Saver Standards for Architectural Design Data, 6th ed., McGraw-Hill, N.Y.)*

CASE 3. - Beam Supported Both Ends - Two Unequal Concentrated Loads, Unequally Distributed.

$$R = \frac{1}{L}\left[W.(L-a) + W_1 b \right]$$

$$R_1 = \frac{1}{L}\left[Wa + W_1 (L-b) \right]$$

V (max.) = Maximum Reaction

At x: when $x > a$ and $< (L-b)$

$$V = R - W$$

At point of load W:

$$M = \frac{a}{L}\left[W (L-a) + W_1 b \right]$$

At point of load W_1:

$$M_1 = \frac{b}{L}\left[Wa + W_1 (L-b) \right]$$

At x: when $x > a$ or $< (L-b)$

$$M = W \frac{a}{L}(L-x) + W_1 \frac{bx}{L}$$

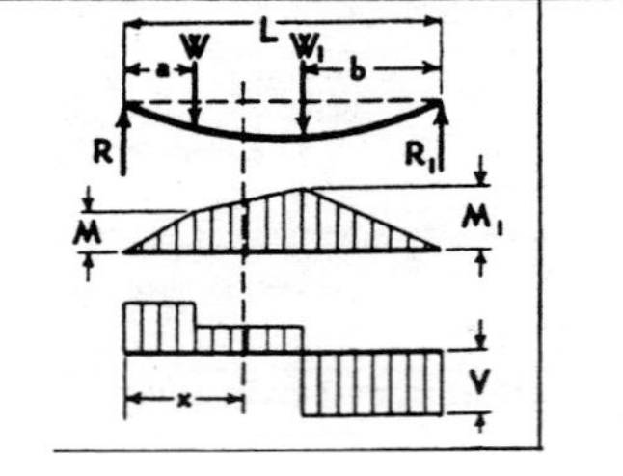

CASE 4. - Beam Supported Both Ends - Three Unequal Concentrated Loads, Unequally Distributed.

$$R = \frac{Wb + W_1 b_1 + W_2 b_2}{L}$$

$$R_1 = \frac{Wa + W_1 a_1 + W_2 a_2}{L}$$

V (max.) = Maximum Reaction

At x: when $x > a$ and $< a_1$

$$V = R - W$$

At x: when $x > a_1$ and $< a_2$

$$V = R - W - W_1$$

At x: when $x = a$

$$M = Ra$$

At x: when $x = a_1$

$$M_1 = Ra_1 - W(a_1 - a)$$

At x: when $x = a_2$

$$M_2 = Ra_2 - W(a_2 - a) - W_1(a_2 - a_1)$$

M (max.) $= M$ when $W = R$ or $> R$

M (max.) $= M_1$ when $\begin{cases} W_1 + W = R \text{ or} > R \\ W_1 + W_2 = R_1 \text{ or} > R_1 \end{cases}$

M (max.) $= M_2$ when $W_2 = R_1$ or $> R_1$

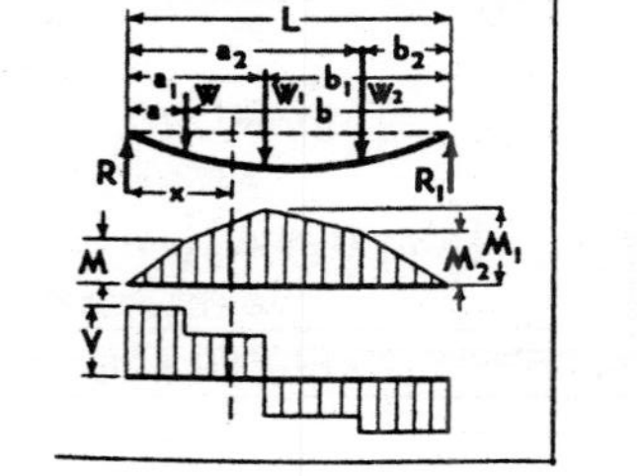

FIGURE 2.2 (*Continued*) Beam formulas.

CASE 5. - Beam Fixed Both Ends - Continuous Load, Uniformly Distributed.

$R = R_1 = V\,(\text{max.}) = \frac{W}{2}$ At x: $V = \frac{W}{2} - \frac{Wx}{L}$	At center: $M\,(\text{max.}) = \frac{WL}{24}$ At supports: $M_1\,(\text{max.}) = \frac{WL}{12}$ At x: $M = \frac{W}{2L}\left(-\frac{L^2}{6} + Lx - x^2\right)$	At center: $D\,(\text{max.}) = \frac{1}{384}\,\frac{WL^3}{EI}$ At x: $D = \frac{Wx^2}{24\,EIL}\left(L^2 - 2Lx + x^2\right)$

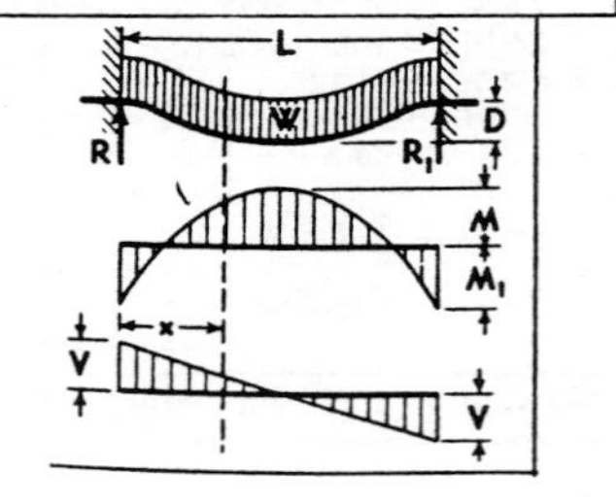

CASE 6. - Beam Fixed Both Ends - Concentrated Load at Any Point.

$$R = W\left(\frac{b^2(3a+b)}{L^3}\right)$$

$$R_1 = W\left(\frac{a^2(3b+a)}{L^3}\right)$$

V (max.) $= R$ when $a < b$

$= R_1$ when $a > b$

At x: when $x < a$

$V = R$

At support R:

M_1 (max. neg. mom. when $b > a$) $= -W\dfrac{ab^2}{L^2}$

At support R_1:

M_2 (max. neg. mom. when $a > b$) $= -W\dfrac{a^2b}{L^2}$

At point of load:

M (max.) $= Ra + M_1 = Ra - W\dfrac{ab^2}{L^2}$

At x: $M = Rx - W\dfrac{ab^2}{L^2}$

At x: when $x = \dfrac{2aL}{3a+b}$ and $a > b$

$$D\,(\text{max.}) = \frac{2Wa^3b^2}{3EI(3a+b)^2}$$

when $x < a$

$$D = \frac{Wb^2x^2}{6EIL^3}(3aL - 3ax - bx)$$

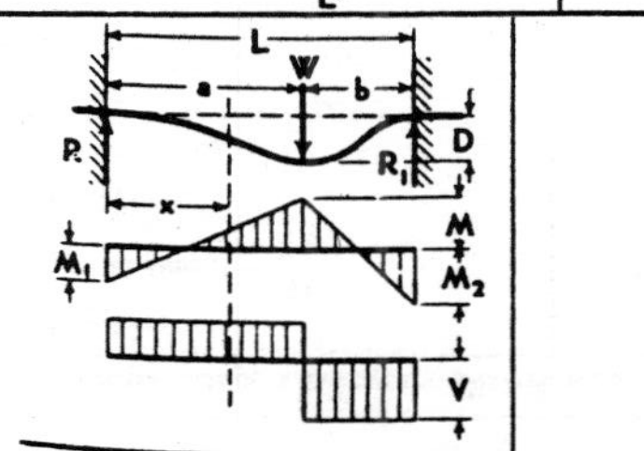

FIGURE 2.2 (*Continued*) Beam formulas.

CASE 7. - Beam Fixed at One End (Cantilever) - Continuous Load, Uniformly Distributed..

$R_1 = V\,(\text{max.}) = W$	At fixed end: $M\,(\text{max.}) = \frac{WL}{2}$	At free end: $D\,(\text{max.}) = \frac{WL^3}{8\,EI}$
At x: $V = \frac{Wx}{L}$	At x: $M = \frac{Wx^2}{2L}$	At x: $D = \frac{W}{24\,EIL}\left(x^4 - 4L^3x + 3L^4\right)$

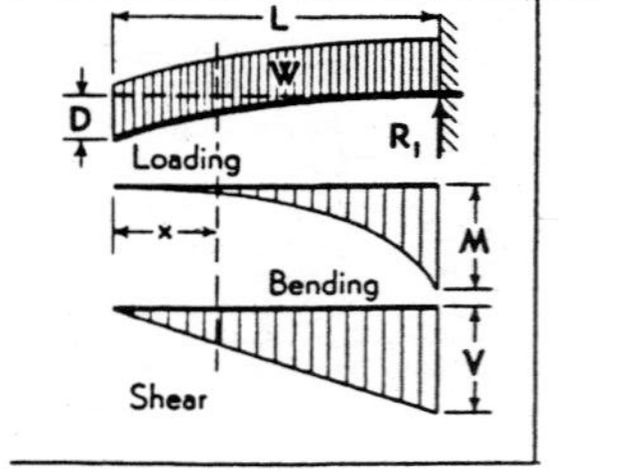

CASE 8. - Beam Fixed at One End (Cantilever) - Concentrated Load at Any Point.

$R_1 = V\,(max.) = W$ At x: when $x > a$ $V = W$ At x: when $x < a$ $V = 0$	At fixed end: $M\,(max.) = Wb$ At x: when $x > a$ $M = W\,(x - a)$	At free end: $D\,(max.) = \frac{WL^3}{6EI}\left[2 - \frac{3a}{L} + \left(\frac{a}{L}\right)^3\right]$ At point of load: $D = \frac{W}{3EI}(L - a)^3$ At x: when $x > a$ $D = \frac{W}{6EI}\left(-3aL^2 + 2L^3 + x^3 - 3ax^2 - 3L^2x + 6aLx\right)$

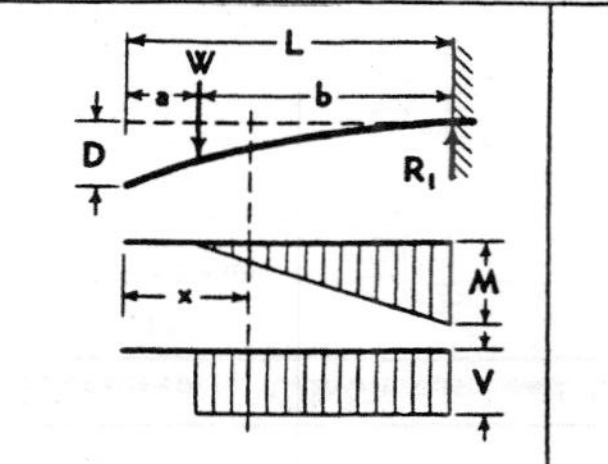

FIGURE 2.2 (*Continued*) Beam formulas.

CASE 9. - Beam Fixed at One End, Supported at Other - Concentrated Load at Any Point.

$$R = W\left(\frac{3b^2L - b^3}{2L^3}\right)$$

$$R_1 = W\left(\frac{3aL^2 - a^3}{2L^3}\right)$$

At x: when $x < a$

$$V = R$$

At x: when $x > a$

$$V = R - W$$

At point of load:

$$M_{(max.)} = Wa\left(\frac{3b^2L - b^3}{2L^3}\right)$$

At fixed end:

$$M_{1\,(max.)} = WL\left(\frac{3b^2L - b^3}{2L^3}\right) - W(L - a)$$

At x: when $x < a$

$$M = Wx\left(\frac{3b^2L - b^3}{2L^3}\right)$$

At x: when $x > a$

$$M = Wx\left(\frac{3b^2L - b^3}{2L^3}\right) - W(x - a)$$

At x: when $x = a = .414L$

$$D_{(max.)} = .0098\frac{WL^3}{EI}$$

At x: when $x < a$

$$D = \frac{1}{6EI}\left[3RL^2x - Rx^3 - 3W(L-a)^2x\right]$$

At x: when $x > a$

$$D = \frac{1}{6EI}\left[R_1(2L^3 - 3L^2x + x^3) - 3Wa(L-x)^2\right]$$

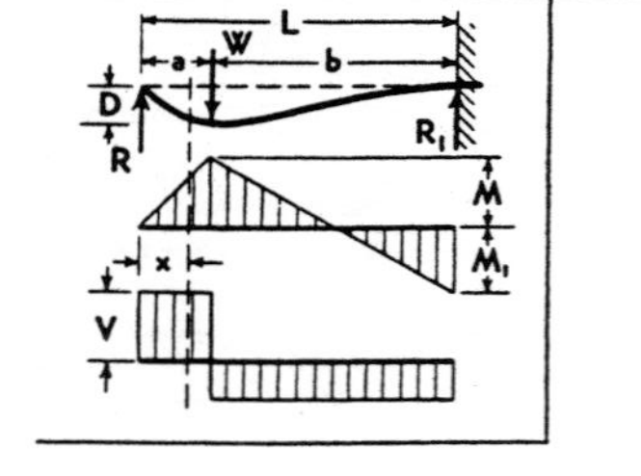

CASE 10. - Beam Fixed at One End, Supported at Other - Continuous Load, Uniformly Distributed.		
$R = \frac{3}{8}W$ $R_1 = V\,(\text{max.}) = \frac{5}{8}W$ At x: $V = \frac{3}{8}W - \frac{Wx}{L}$	At x: when $x = \frac{3}{8}L$ $M\,(\text{max.}) = \frac{9}{128}WL$ At fixed end: $M_1\,(\text{max.}) = \frac{1}{8}WL$ At x: $M = \frac{Wx}{L}\left(\frac{3}{8}L - \frac{1}{2}x\right)$	At x: when $x = .4215\,L$ $D\,(\text{max.}) = .0054\,\frac{WL^3}{EI}$ At x: $D = \frac{Wx}{48\,EIL}\left[-3Lx^2 + 2x^3 + L^3\right]$

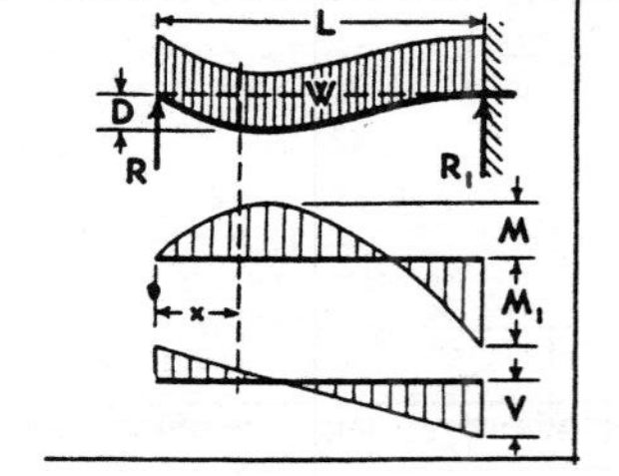

FIGURE 2.2 (*Continued*) Beam formulas.

CASE II. - Beam Overhanging Both Supports, Unsymmetrically Placed - Continuous Load, Uniformly Distributed.

$\frac{W}{a+L+b} = w =$ load per unit of length

$R = w\left[(a+L)^2 - b^2\right] \div 2L$

$R_1 = w\left[(b+L)^2 - a^2\right] \div 2L$

$V\,(\text{max.}) = wa$ or $R - wa$

At x: when $x < a$ $\quad V = w\,(a-x)$

At x_1: when $x_1 < L$ $\quad V = R - w\,(a+x_1)$

At x_2: when $x_2 < b$ $\quad V = w\,(b-x_2)$

At x_1: when $x_1 = \frac{R}{w} - a$

$M\,(\text{max.}) = R\left(\frac{R}{2w} - a\right)$

At R: $\quad M_1 = \frac{1}{2} wa^2$

At R_1: $\quad M_1 = \frac{1}{2} wb^2$

At x: when $x < a$ $\quad M = \frac{1}{2} w\,(a-x)^2$

At x_1: when $x_1 < L$ $\quad M = \frac{1}{2} w\,(a+x_1)^2 - Rx_1$

At x_2: when $x_2 < b$ $\quad M = \frac{1}{2} w\,(b-x_2)^2$

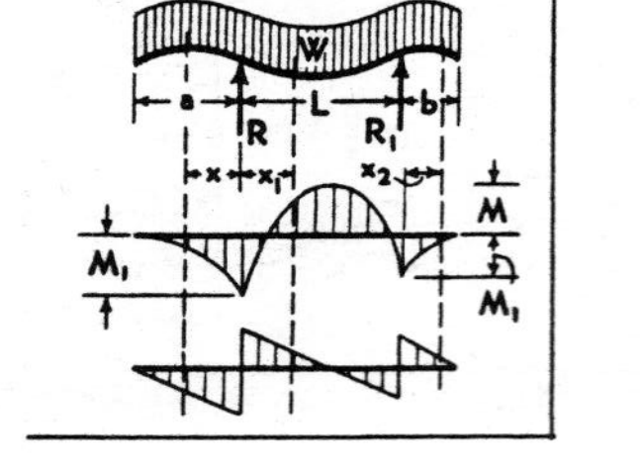

CASE 12. - Beam Overhanging Both Supports, Symmetrically Placed - Two Equal Concentrated Loads at Ends.

$R = R_1 = V\,(\text{max.}) = \frac{W}{2}$ At x: when $x < a$ $V = \frac{W}{2}$	At x_1: when $x_1 < L$ $M\,(\text{max.}) = \frac{Wa}{2}$ At x: when $x < a$ $M = \frac{W}{2}(a - x)$	At free ends: $D = \frac{Wa^2(3L + 2a)}{12EI}$ At center: $D = \frac{WaL^2}{16EI}$

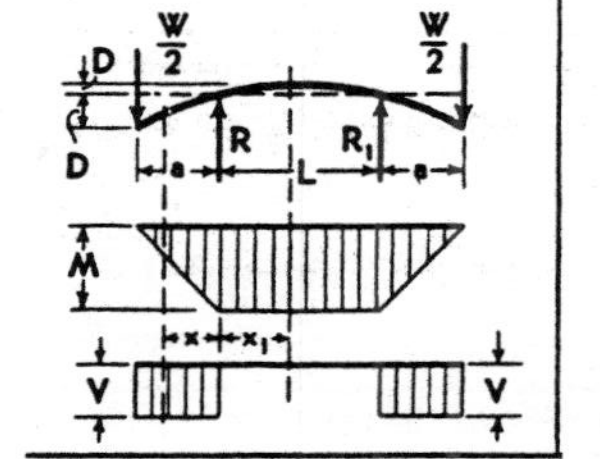

FIGURE 2.2 (*Continued*) Beam formulas.

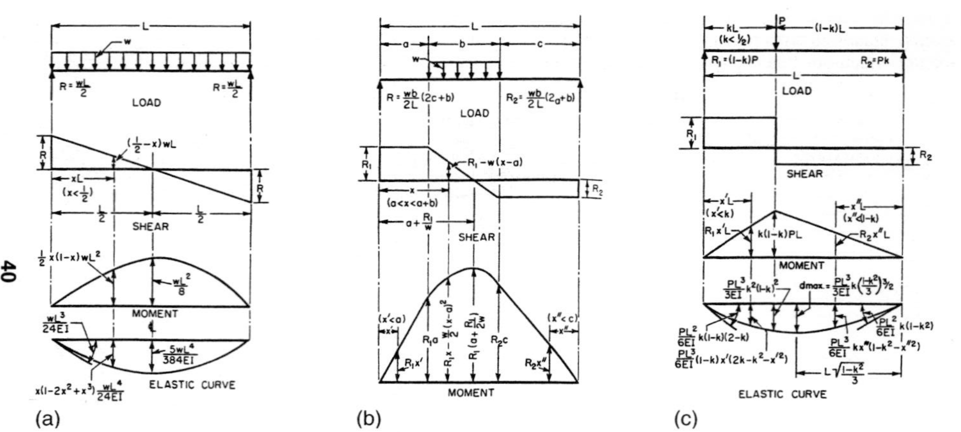

FIGURE 2.3 Elastic-curve equations for prismatic beams. (*a*) Shears, moments, deflections for full uniform load on a simply supported prismatic beam. (*b*) Shears and moments for uniform load over part of a simply supported prismatic beam. (*c*) Shears, moments, deflections for a concentrated load at any point of a simply supported prismatic beam.

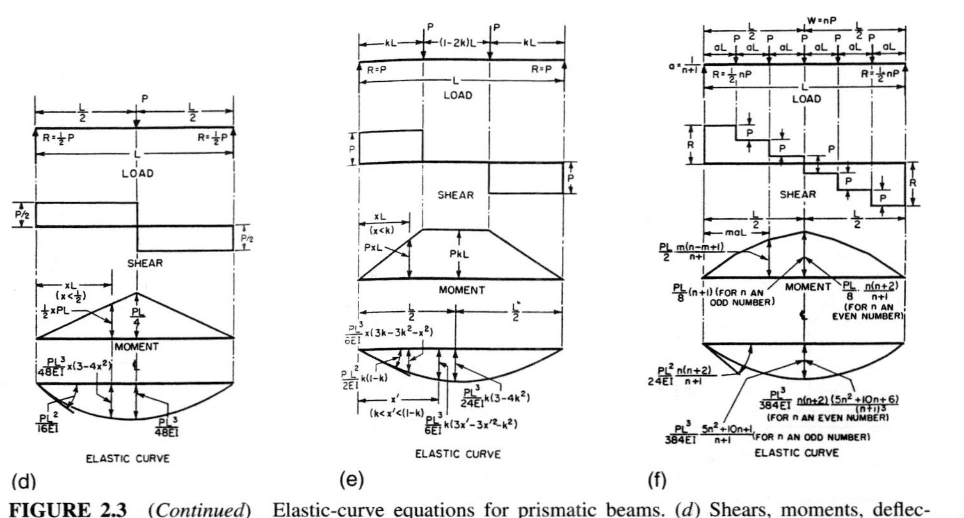

FIGURE 2.3 (*Continued*) Elastic-curve equations for prismatic beams. (*d*) Shears, moments, deflections for a concentrated load at midspan of a simply supported prismatic beam. (*e*) Shears, moments, deflections for two equal concentrated loads on a simply supported prismatic beam. (*f*) Shears, moments, deflections for several equal loads equally spaced on a simply supported prismatic beam.

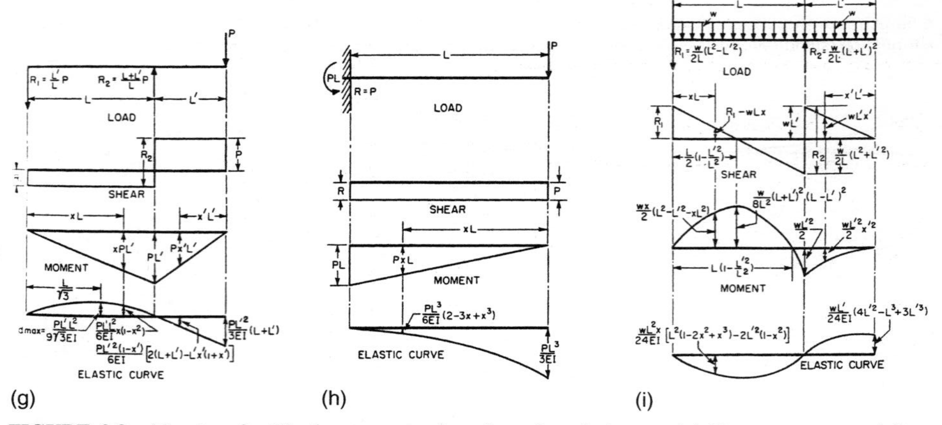

FIGURE 2.3 (*Continued*) Elastic-curve equations for prismatic beams. (*g*) Shears, moments, deflections for a concentrated load on a beam overhang. (*h*) Shears, moments, deflections for a concentrated load on the end of a prismatic cantilever. (*i*) Shears, moments, deflections for a uniform load over the full length of a beam with overhang.

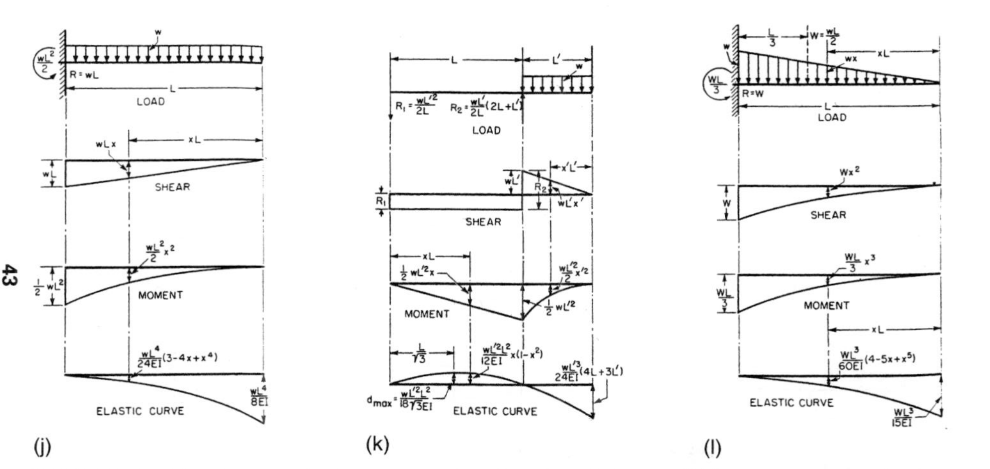

FIGURE 2.3 (*Continued*) Elastic-curve equations for prismatic beams. (*j*) Shears, moments, deflections for uniform load over the full length of a cantilever. (*k*) Shears, moments, deflections for uniform load on a beam overhang. (*l*) Shears, moments, deflections for triangular loading on a prismatic cantilever.

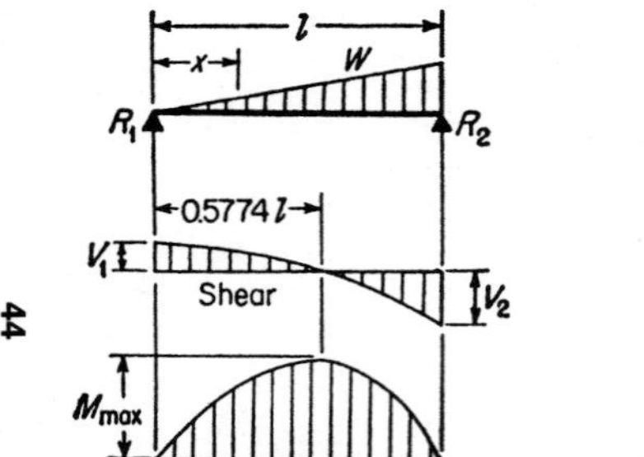

$$R_1 = V_1 \qquad = \frac{W}{3}$$

$$R_2 = V_{2\max} \qquad = \frac{2W}{3}$$

$$V_x \qquad = \frac{W}{3} - \frac{Wx^2}{l^2}$$

$$M_{\max}\left(\text{at } x = \frac{l}{\sqrt{3}} = .5774l\right) \qquad = \frac{2Wl}{9\sqrt{3}} = .1283Wl$$

$$M_x \qquad = \frac{Wx}{3l^2}(l^2 - x^2)$$

$$\Delta_{\max}\left(\text{at } x = l\sqrt{1 - \sqrt{\frac{8}{15}}} = .5193l\right) = .01304\frac{Wl^3}{EI}$$

$$\Delta_x \qquad = \frac{Wx}{180EIl^2}(3x^4 - 10l^2x^2 + 7l^4)$$

(m)

FIGURE 2.3 (*Continued*) Elastic-curve equations for prismatic beams. (*m*) Simple beam—load increasing uniformly to one end.

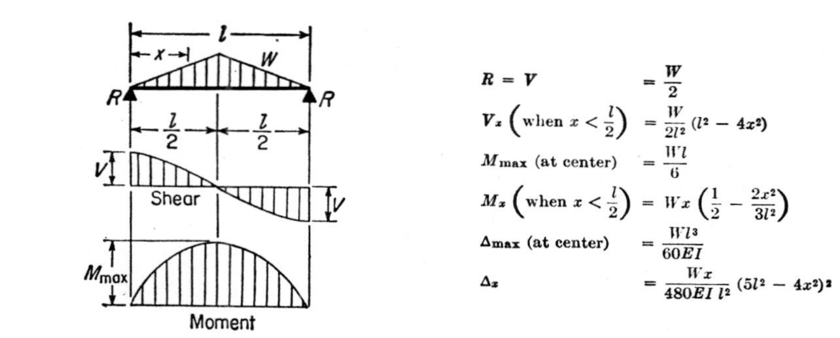

$$R = V = \frac{W}{2}$$

$$V_x \left(\text{when } x < \frac{l}{2}\right) = \frac{W}{2l^2}(l^2 - 4x^2)$$

$$M_{\max} \text{ (at center)} = \frac{Wl}{6}$$

$$M_x \left(\text{when } x < \frac{l}{2}\right) = Wx\left(\frac{1}{2} - \frac{2x^2}{3l^2}\right)$$

$$\Delta_{\max} \text{ (at center)} = \frac{Wl^3}{60EI}$$

$$\Delta_x = \frac{Wx}{480EI\,l^2}(5l^2 - 4x^2)^2$$

FIGURE 2.3 (*Continued*) Elastic-curve equations for prismatic beams. (*n*) Simple beam—load increasing uniformly to center.

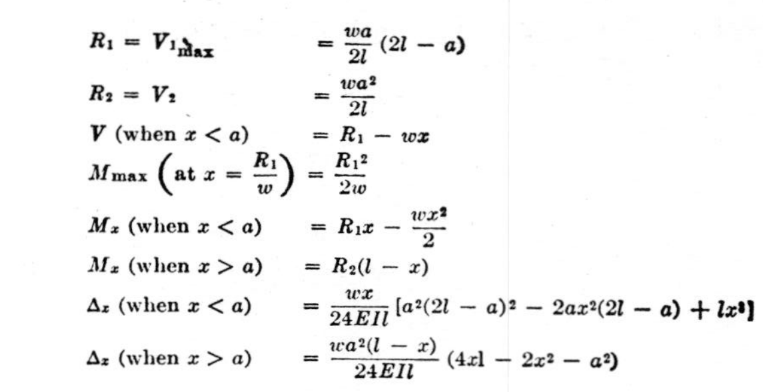

$$R_1 = V_{1\max} = \frac{wa}{2l}(2l - a)$$

$$R_2 = V_2 = \frac{wa^2}{2l}$$

$$V \text{ (when } x < a) = R_1 - wx$$

$$M_{\max}\left(\text{at } x = \frac{R_1}{w}\right) = \frac{R_1^2}{2w}$$

$$M_x \text{ (when } x < a) = R_1 x - \frac{wx^2}{2}$$

$$M_x \text{ (when } x > a) = R_2(l - x)$$

$$\Delta_x \text{ (when } x < a) = \frac{wx}{24EIl}[a^2(2l - a)^2 - 2ax^2(2l - a) + lx^3]$$

$$\Delta_x \text{ (when } x > a) = \frac{wa^2(l - x)}{24EIl}(4xl - 2x^2 - a^2)$$

(o)

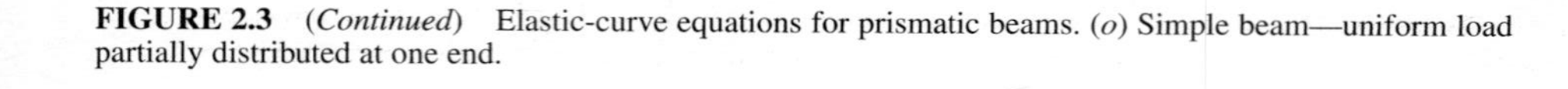

FIGURE 2.3 (*Continued*) Elastic-curve equations for prismatic beams. (*o*) Simple beam—uniform load partially distributed at one end.

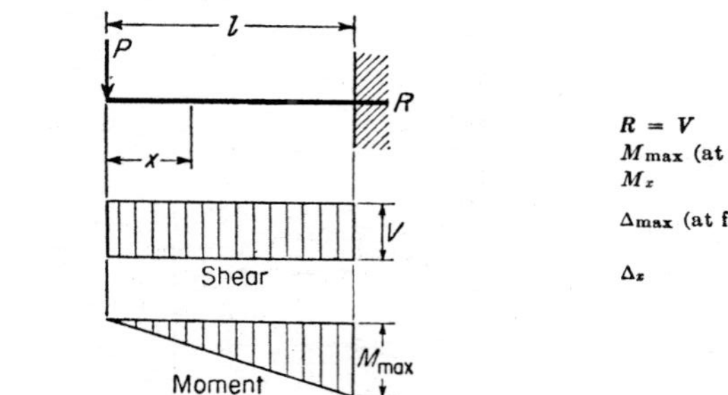

$$R = V = P$$
$$M_{max} \text{ (at fixed end)} = Pl$$
$$M_x = Px$$
$$\Delta_{max} \text{ (at free end)} = \frac{Pl^3}{3EI}$$
$$\Delta_x = \frac{P}{6EI}(2l^3 - 3l^2x + x^3)$$

FIGURE 2.3 (*Continued*) Elastic-curve equations for prismatic beams. (*p*) Cantilever beam—concentrated load at free end.

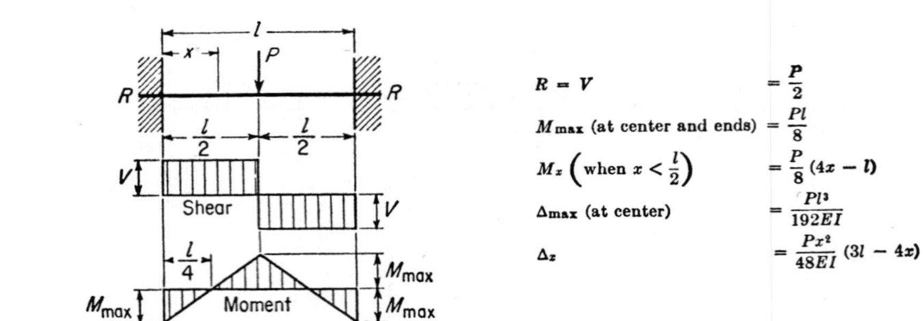

$$R = V = \frac{P}{2}$$

$$M_{\max} \text{ (at center and ends)} = \frac{Pl}{8}$$

$$M_x \left(\text{when } x < \frac{l}{2}\right) = \frac{P}{8}(4x - l)$$

$$\Delta_{\max} \text{ (at center)} = \frac{Pl^3}{192EI}$$

$$\Delta_x = \frac{Px^2}{48EI}(3l - 4x)$$

FIGURE 2.3 (*Continued*) Elastic-curve equations for prismatic beams. (*q*) Beam fixed at both ends—concentrated load at center.

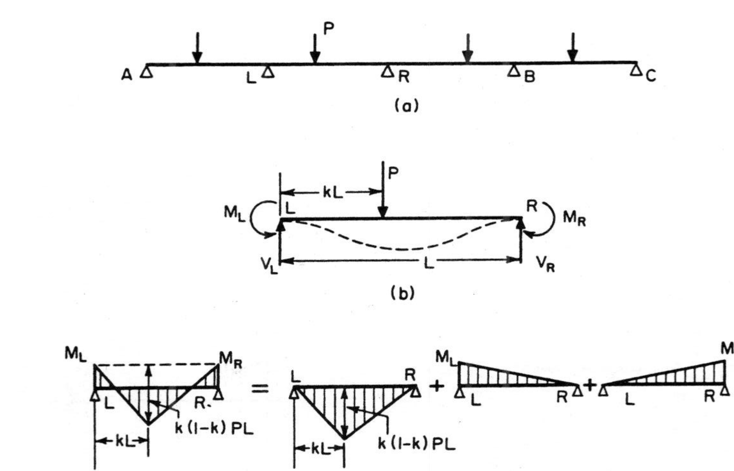

FIGURE 2.4 Any span of a continuous beam (*a*) can be treated as a simple beam, as shown in (*b*) and (*c*). In (*c*), the moment diagram is decomposed into basic components.

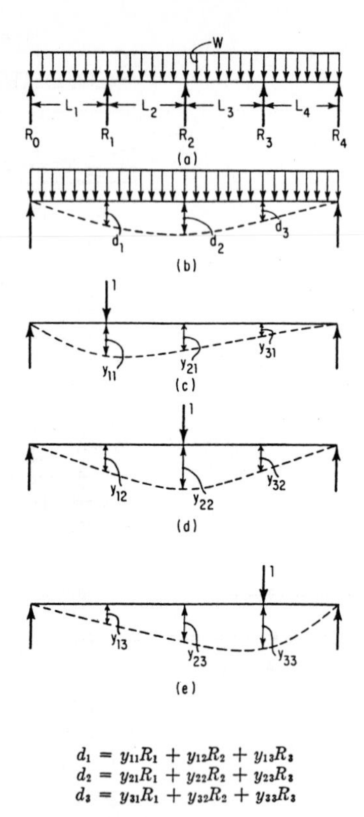

$$d_1 = y_{11}R_1 + y_{12}R_2 + y_{13}R_3$$
$$d_2 = y_{21}R_1 + y_{22}R_2 + y_{23}R_3$$
$$d_3 = y_{31}R_1 + y_{32}R_2 + y_{33}R_3$$

FIGURE 2.5 Reactions of continuous beam (*a*) found by making the beam statically determinate. (*b*) Deflections computed with interior supports removed. (*c*), (*d*), and (*e*) Deflections calculated for unit load over each removed support, to obtain equations for each redundant.

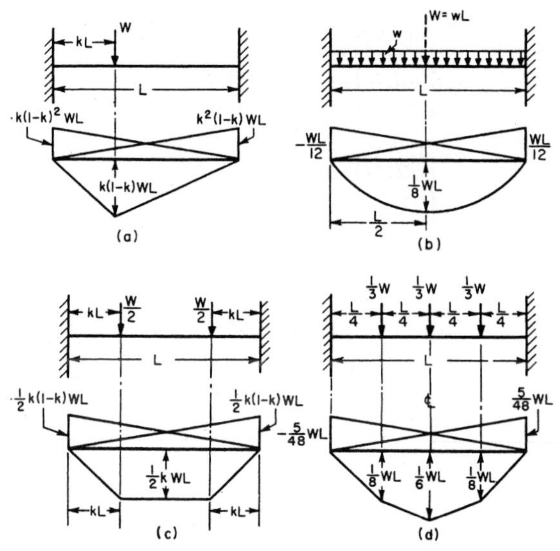

FIGURE 2.6 Fixed-end moments for a prismatic beam. (*a*) For a concentrated load. (*b*) For a uniform load. (*c*) For two equal concentrated loads. (*d*) For three equal concentrated loads.

In a similar manner the fixed-end moment for a beam with one end hinged and the supports at different levels can be found from

$$M^F = K\frac{d}{L} \tag{2.3}$$

where K is the actual stiffness for the end of the beam that is fixed; for beams of variable moment of inertia K equals the fixed-end stiffness times $(1 - C_L^F C_R^F)$.

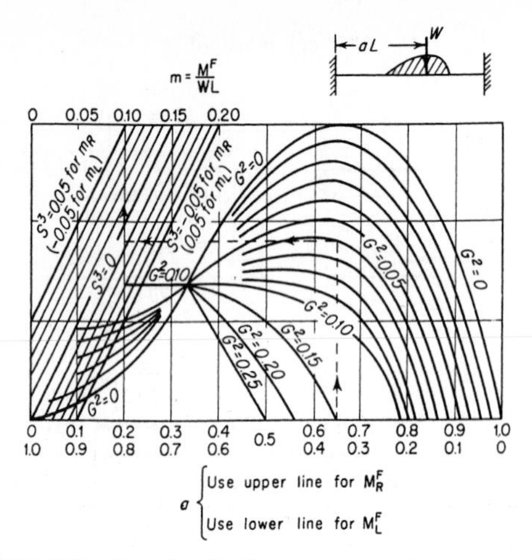

FIGURE 2.7 Chart for fixed-end moments due to any type of loading.

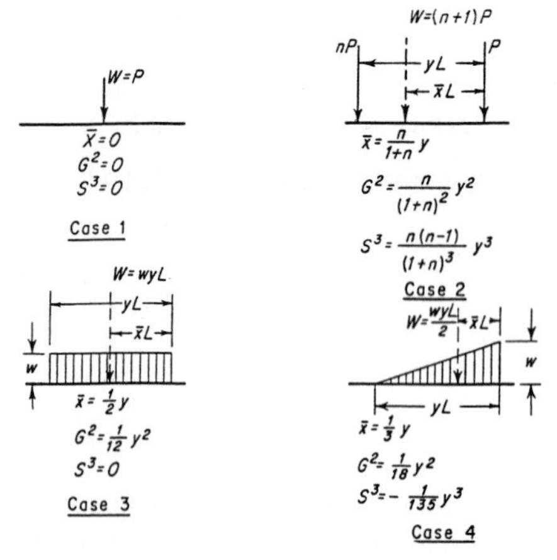

FIGURE 2.8 Characteristics of loadings.

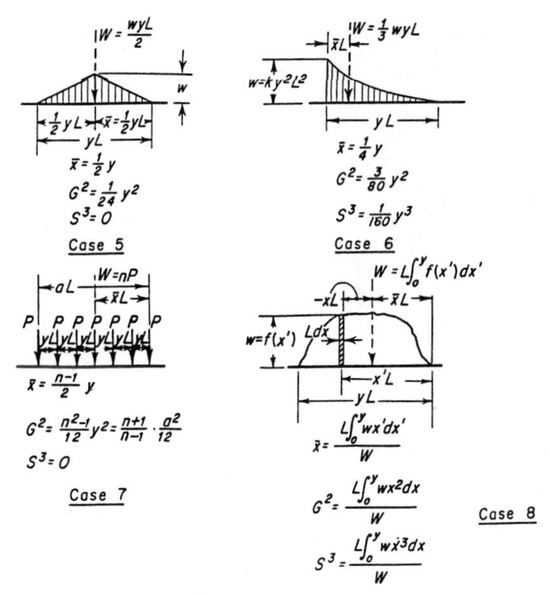

FIGURE 2.8 (*Continued*) Characteristics of loadings.

ULTIMATE STRENGTH OF CONTINUOUS BEAMS

Methods for computing the ultimate strength of continuous beams and frames may be based on two theorems that fix upper and lower limits for load-carrying capacity:

1. *Upper-bound theorem.* A load computed on the basis of an assumed link mechanism is always greater than, or at best equal to, the ultimate load.

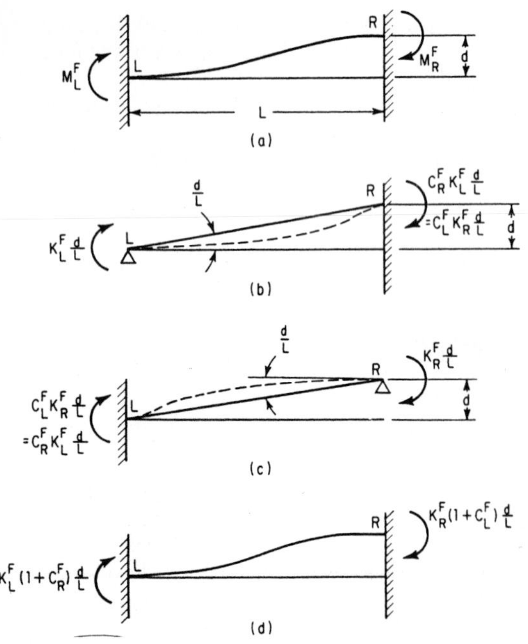

FIGURE 2.9 Moments due to deflection of a fixed-end beam.

2. *Lower-bound theorem.* The load corresponding to an equilibrium condition with arbitrarily assumed values for the redundants is smaller than, or at best equal to, the ultimate loading—provided that everywhere moments do not exceed M_P. The equilibrium method, based on the lower bound theorem, is usually easier for simple cases.

For the continuous beam in Fig. 2.10, the ratio of the plastic moment for the end spans is k times that for the center span ($k > 1$).

Figure 2.10*b* shows the moment diagram for the beam made determinate by ignoring the moments at B and C and the moment diagram for end moments M_B and M_C applied to the determinate beam. Then, by using Fig. 2.10*c*, equilibrium is maintained when

$$M_P = \frac{wL^2}{4} - \frac{1}{2}M_B - \frac{1}{2}M_C$$

$$= \frac{wL^2}{4 - kM_P}$$

$$= \frac{wL^2}{4(1 + k)} \tag{2.4}$$

The mechanism method can be used to analyze rigid frames of constant section with fixed bases, as in Fig. 2.11. Using this method with the vertical load at midspan equal to 1.5 times the lateral load, the ultimate load for the frame is $4.8M_P/L$ laterally and $7.2M_P/L$ vertically at midspan.

Maximum moment occurs in the interior spans AB and CD when

$$x = \frac{L}{2} - \frac{M}{wL} \tag{2.5}$$

or if

$$M = kM_P \qquad \text{when} \qquad x = \frac{L}{2} - \frac{kM_P}{wL} \tag{2.6}$$

A plastic hinge forms at this point when the moment equals kM_P. For equilibrium,

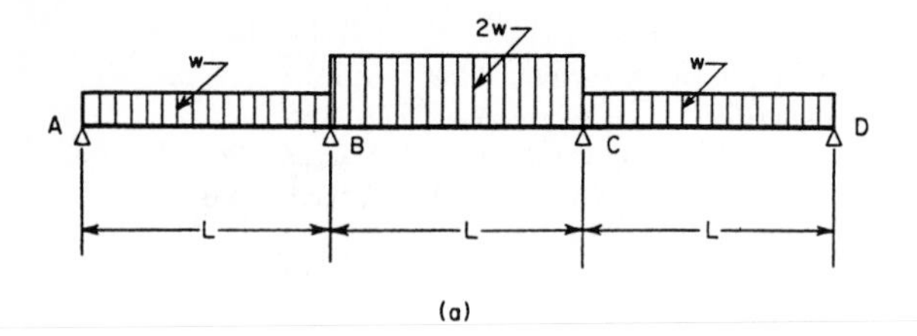

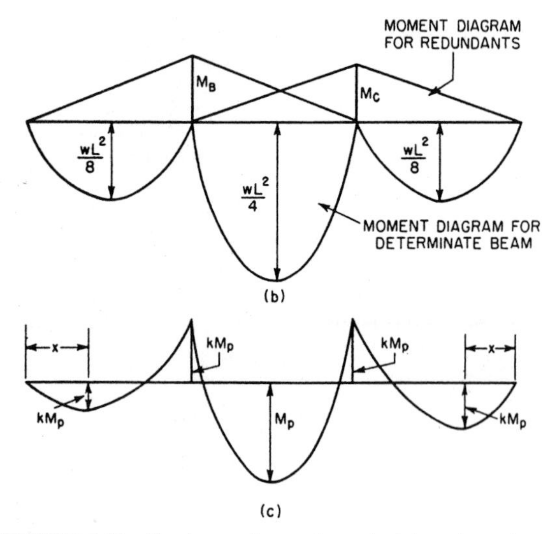

FIGURE 2.10 Continuous beam shown in (*a*) carries twice as much uniform load in the center span as in the side span. In (*b*) are shown the moment diagrams for this loading condition with redundants removed and for the redundants. The two moment diagrams are combined in (*c*), producing peaks at which plastic hinges are assumed to form.

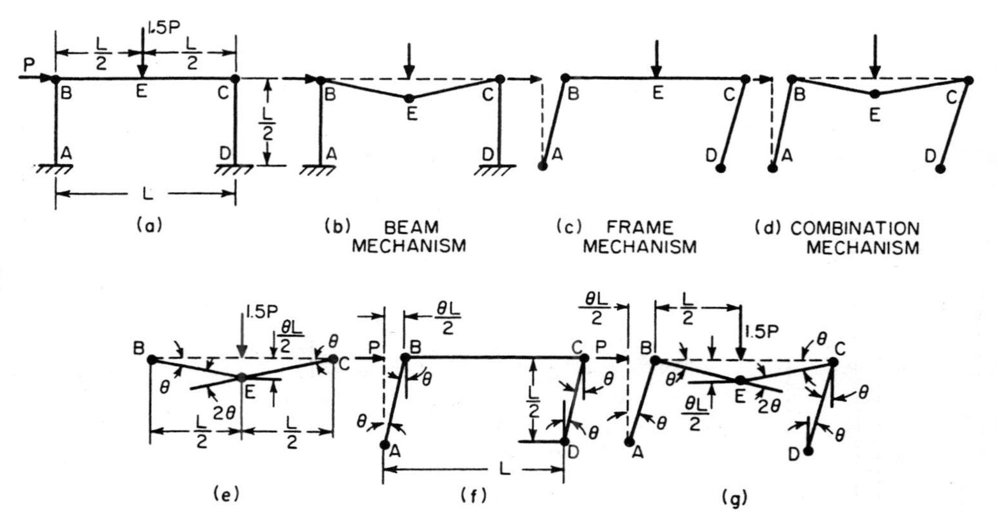

FIGURE 2.11 Ultimate-load possibilities for a rigid frame of constant section with fixed bases.

$$kM_P = \frac{w}{2} x (L - x) - \frac{x}{L} kM_P$$

$$= \frac{w}{2}\left(\frac{L}{2} - \frac{kM_P}{wL}\right)\left(\frac{L}{2} + \frac{KM_P}{wL}\right) - \left(\frac{1}{2} - \frac{kM_P}{wL^2}\right) kM_P$$

leading to

$$\frac{k^2M_P^2}{wL^2} - 3kM_P + \frac{wL^2}{4} = 0 \qquad (2.7)$$

When the value of M_P previously computed is substituted,

$$7k^2 + 4k = 4 \qquad \text{or} \qquad k(k + 4/7) = 4/7$$

from which $k = 0.523$. The ultimate load is

$$wL = \frac{4M_P(1 + k)}{L} = 6.1\frac{M_P}{L} \qquad (2.8)$$

In any continuous beam, the bending moment at any section is equal to the bending moment at any other section, plus the shear at that section times its arm, plus the product of all the intervening external forces times their respective arms. Thus, in Fig. 2.12,

$$V_x = R_1 + R_2 + R_3 - P_1 - P_2 - P_3$$

$$M_x = R_1(l_1 + l_2 + x) + R_2(l_2 + x) + R_3x$$

$$- P_1(l_2 + c + x) - P_2(b + x) - P_3a$$

$$M_x = M_3 + V_3x - P_3a$$

Table 2.1 gives the value of the moment at the various supports of a uniformly loaded continuous beam over equal

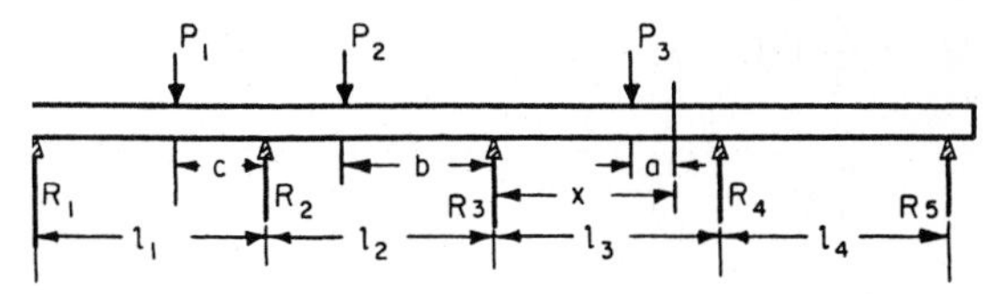

FIGURE 2.12 Continuous beam.

spans, and it also gives the values of the shears on each side of the supports. Note that the shear is of the opposite sign on either side of the supports and that the sum of the two shears is equal to the reaction.

Figure 2.13 shows the relation between the moment and shear diagrams for a uniformly loaded continuous beam of four equal spans. (See Table 2.1.) Table 2.1 also gives the maximum bending moment that occurs between supports, in addition to the position of this moment and the points of

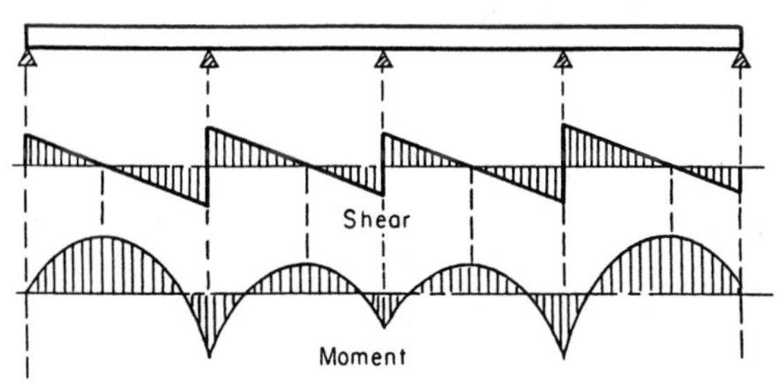

FIGURE 2.13 Relation between moment and shear diagrams for a uniformly loaded continuous beam of four equal spans.

TABLE 2.1 Uniformly Loaded Continuous Beams over Equal Spans

(Uniform load per unit length = w; length of each span = l)

Number of supports	Notation of support of span	Shear on each side of support. L = left, R = right. Reaction at any support is $L + R$		Moment over each support	Max moment in each span	Distance to point of max moment, measured to right from from support	Distance to point of inflection, measured to right from support
		L	R				
2	1 or 2	0	$^1/_2$	0	0.125	0.500	None
3	1	0	$^3/_8$	0	0.0703	0.375	0.750
	2	$^5/_8$	$^5/_8$	$^1/_8$	0.0703	0.625	0.250
4	1	0	$^4/_{10}$	0	0.080	0.400	0.800
	2	$^6/_{10}$	$^5/_{10}$	$^1/_{10}$	0.025	0.500	0.276, 0.724
	1	0	$^{11}/_{28}$	0	0.0772	0.393	0.786
5	2	$^{17}/_{28}$	$^{15}/_{28}$	$^3/_{28}$	0.0364	0.536	0.266, 0.806
	3	$^{13}/_{28}$	$^{13}/_{28}$	$^2/_{28}$	0.0364	0.464	0.194, 0.734
	1	0	$^{15}/_{38}$	0	0.0779	0.395	0.789

6	2	$^{23}/_{38}$	$^{20}/_{38}$	$^{4}/_{38}$	0.0332	0.526	0.268, 0.783
	3	$^{18}/_{38}$	$^{19}/_{38}$	$^{3}/_{38}$	0.0461	0.500	0.196, 0.804
	1	0	$^{41}/_{104}$	0	0.0777	0.394	0.788
	2	$^{63}/_{104}$	$^{55}/_{104}$	$^{11}/_{104}$	0.0340	0.533	0.268, 0.790
7	3	$^{49}/_{104}$	$^{51}/_{104}$	$^{8}/_{104}$	0.0433	0.490	0.196, 0.785
	4	$^{53}/_{104}$	$^{53}/_{104}$	$^{9}/_{104}$	0.0433	0.510	0.215, 0.804
	1	0	$^{56}/_{142}$	0	0.0778	0.394	0.789
8	2	$^{86}/_{142}$	$^{75}/_{142}$	$^{15}/_{142}$	0.0338	0.528	0.268, 0.788
	3	$^{67}/_{142}$	$^{70}/_{142}$	$^{11}/_{142}$	0.0440	0.493	0.196, 0.790
	4	$^{72}/_{142}$	$^{71}/_{142}$	$^{12}/_{142}$	0.0405	0.500	0.215, 0.785
Values apply to		w^1	w^1	wl^2	wl^2	l	l

The numerical values given are coefficients of the expressions at the foot of each column.

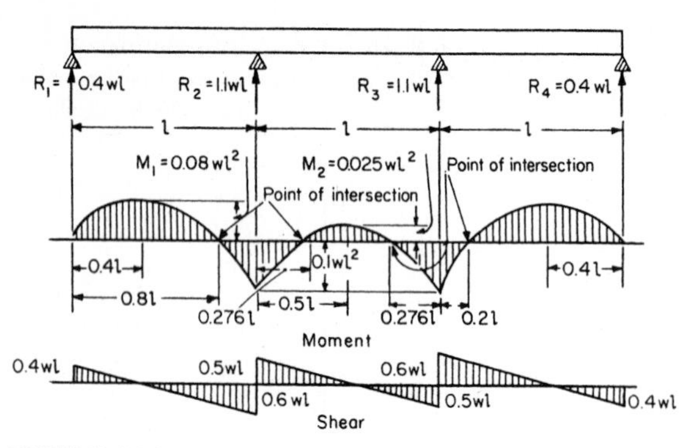

FIGURE 2.14 Values of the functions for a uniformly loaded continuous beam resting on three equal spans with four supports.

inflection. Figure 2.14 shows the values of the functions for a uniformly loaded continuous beam resting on three equal spans with four supports.

Maxwell's Theorem

When a number of loads rest upon a beam, the deflection at any point is equal to the sum of the deflections at this point due to each of the loads taken separately. Maxwell's theorem states that if unit loads rest upon a beam at two points, A and B, the deflection at A due to the unit load at B equals the deflection at B due to the unit load at A.

Castigliano's Theorem

This theorem states that the deflection of the point of application of an external force acting on a beam is equal to the partial derivative of the work of deformation with respect to this force. Thus, if P is the force, f is the deflection, and U is the work of deformation, which equals the resilience:

$$\frac{dU}{dP} = f$$

According to the *principle of least work,* the deformation of any structure takes place in such a manner that the work of deformation is a minimum.

BEAMS OF UNIFORM STRENGTH

Beams of uniform strength so vary in section that the unit stress S remains constant, and I/c varies as M. For rectangular beams of breadth b and depth d, $I/c = I/c = bd^2/6$ and $M = Sbd^2/6$. For a cantilever beam of rectangular cross section, under a load P, $Px = Sbd^2/6$. If b is constant, d^2 varies with x, and the profile of the shape of the beam is a parabola, as in Fig. 2.15. If d is constant, b varies as x, and the beam is triangular in plan (Fig. 2.16).

Shear at the end of a beam necessitates modification of the forms determined earlier. The area required to resist shear is P/S_v in a cantilever and R/S_v in a simple beam. Dotted extensions in Figs. 2.15 and 2.16 show the changes necessary to enable these cantilevers to resist shear. The waste in material and extra cost in fabricating, however, make many of the forms impractical, except for cast iron. Figure 2.17 shows some of the simple sections of uniform strength. In none of these, however, is shear taken into account.

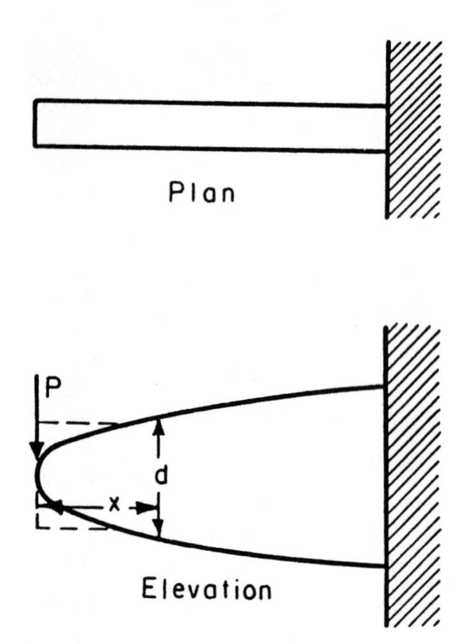

FIGURE 2.15 Parabolic beam of uniform strength.

SAFE LOADS FOR BEAMS OF VARIOUS TYPES

Table 2.2 gives 32 formulas for computing the approximate safe loads on steel beams of various cross sections for an allowable stress of 16,000 lb/in^2 (110.3 MPa). Use these formulas for quick estimation of the safe load for any steel beam you are using in a design.

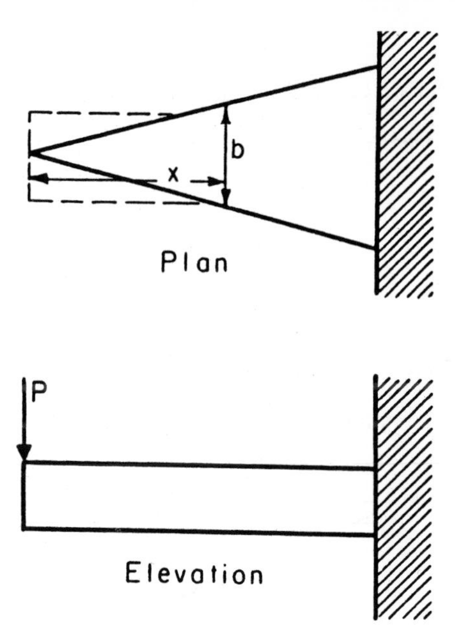

FIGURE 2.16 Triangular beam of uniform strength.

Table 2.3 gives coefficients for correcting values in Table 2.2 for various methods of support and loading. When combined with Table 2.2, the two sets of formulas provide useful time-saving means of making quick safe-load computations in both the office and the field.

TABLE 2.2 Approximate Safe Loads in Pounds (kgf) on Steel Beams*

(Beams supported at both ends; allowable fiber stress for steel, 16,000 lb/in² (1.127 kgf/cm²) (basis of table) for iron, reduce values given in table by one-eighth)

Shape of section	Greatest safe load, lb‡		Deflection, in‡	
	Load in middle	Load distributed	Load in middle	Load distributed
Solid rectangle	$\frac{890AD}{L}$	$\frac{1,780AD}{L}$	$\frac{wL^3}{32AD^2}$	$\frac{wL^3}{52AD^2}$
Hollow rectangle	$\frac{890(AD - ad)}{L}$	$\frac{1,780(AD - ad)}{L}$	$\frac{wL^3}{32(AD^2 - ad^2)}$	$\frac{wL^3}{52(AD^2 - ad^2)}$
Solid cylinder	$\frac{667AD}{L}$	$\frac{1,333AD}{L}$	$\frac{wL^3}{24AD^2}$	$\frac{wL^3}{38AD^2}$
Hollow cylinder	$\frac{667(AD - ad)}{L}$	$\frac{1,333(AD - ad)}{L}$	$\frac{wL^3}{24(AD^2 - ad^2)}$	$\frac{wL^3}{38(AD^2 - ad^2)}$

Even-legged angle or tee	$\frac{885AD}{L}$	$\frac{1.770AD}{L}$	$\frac{wL^3}{32AD^2}$	$\frac{wL^3}{52AD^2}$
Channel or Z bar	$\frac{1,525AD}{L}$	$\frac{3,050AD}{L}$	$\frac{wL^3}{53AD^2}$	$\frac{wL^3}{85AD^2}$
Deck beam	$\frac{1,380AD}{L}$	$\frac{2,760AD}{L}$	$\frac{wL^3}{50AD^2}$	$\frac{wL^3}{80AD^2}$
I beam	$\frac{1,795AD}{L}$	$\frac{3,390AD}{L}$	$\frac{wL^3}{58AD^2}$	$\frac{wL^3}{93AD^2}$

*L = distance between supports, ft (m); A = sectional area of beam, in^2 (cm^2); D = depth of beam, in (cm); a = interior area, in^2 (cm^2); d = interior depth, in (cm); w = total working load, net tons (kgf).

TABLE 2.3 Coefficients for Correcting Values in Table 2.2 for Various Methods of Support and of Loading†

Conditions of loading	Max relative safe load	Max relative deflection under max relative safe load
Beam supported at ends		
Load uniformly distributed over span	1.0	1.0
Load concentrated at center of span	$\frac{1}{2}$	0.80
Two equal loads symmetrically concentrated	$l/4c$	
Load increasing uniformly to one end	0.974	0.976
Load increasing uniformly to center	$\frac{3}{4}$	0.96
Load decreasing uniformly to center	$\frac{3}{2}$	1.08

Beam fixed at one end, cantilever		
Load uniformly distributed over span	$1/4$	2.40
Load concentrated at end	$1/8$	3.20
Load increasing uniformly to fixed end	$3/8$	1.92
Beam continuous over two supports equidistant from ends		
Load uniformly distributed over span		
1. If distance $a > 0.2071l$	$l^2/4a^2$	
2. If distance $a < 0.2071l$	$l/(l-4a)$	
3. If distance $a = 0.2071l$	5.83	
Two equal loads concentrated at ends	$l/4a$	

† l = length of beam; c = distance from support to nearest concentrated load; a = distance from support to end of beam.

1. Fixed at One End, Load P Concentrated at Other End

Beam	Cross section	Elevation and plan	Formulas
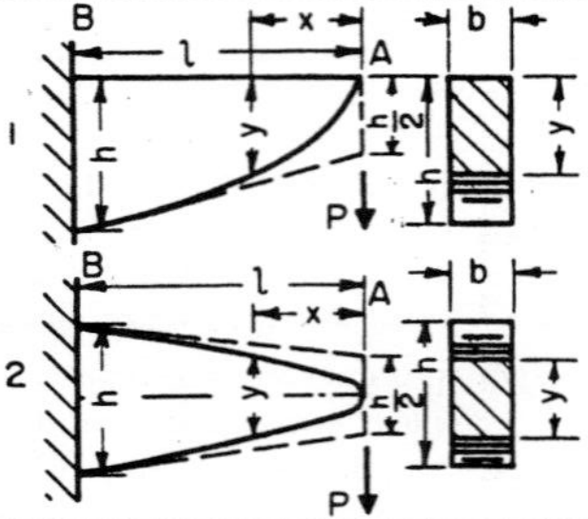 1 2	Rectangle: width (b) constant, depth (y) variable	Elevation: 1, top, straight line; bottom, parabola. 2, complete parabola Plan: rectangle	$y^2 = \frac{6P}{bS_s} x$ $h = \sqrt{\frac{6Pl}{bS_s}}$ Deflection at A: $f = \frac{8P}{bE} \left(\frac{l}{h}\right)^3$

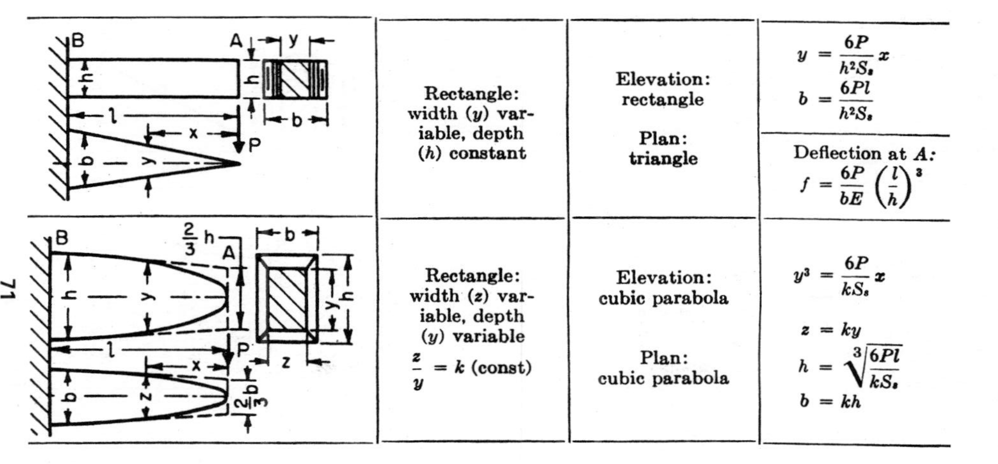

Rectangle: width (y) variable, depth (h) constant	Elevation: rectangle Plan: triangle	$y = \frac{6P}{h^2 S_s} x$ $b = \frac{6Pl}{h^2 S_s}$ Deflection at A: $f = \frac{6P}{bE} \left(\frac{l}{h}\right)^3$
Rectangle: width (z) variable, depth (y) variable $\frac{z}{y} = k$ (const)	Elevation: cubic parabola Plan: cubic parabola	$y^3 = \frac{6P}{kS_s} x$ $z = ky$ $h = \sqrt[3]{\frac{6Pl}{kS_s}}$ $b = kh$

FIGURE 2.17 Beams of uniform strength (in bending).

Beam	Cross section	Elevation and plan	Formulas
1. Fixed at One End, Load P Concentrated at Other End (cont.)			
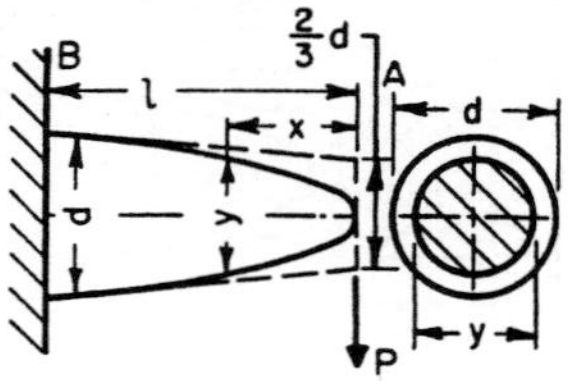	Circle: diam (y) variable	Elevation: cubic parabola Plan: cubic parabola	$y^3 = \frac{32P}{\pi S_s} x$ $d = \sqrt[3]{\frac{32Pl}{\pi S_s}}$
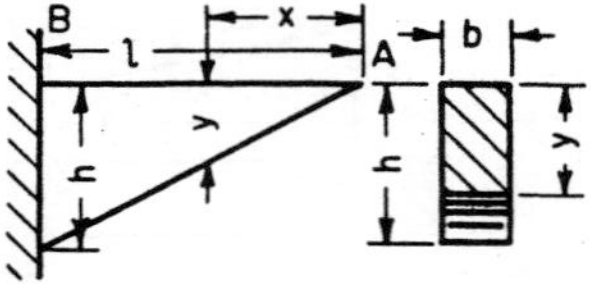	Rectangle: width (b) constant, depth (y) variable	Elevation: triangle Plan: rectangle	$y = x \sqrt{\frac{3P}{blS}}$ $h = \sqrt{\frac{3Pl}{bS_s}}$ $f = 6 \frac{P}{bE} \left(\frac{l}{h}\right)^3$

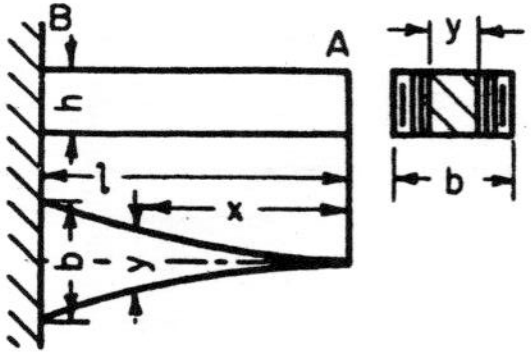	Rectangle: width (y) variable, depth (h) constant	Elevation: rectangle Plan: two parabolic curves with vertices at free end	$y = \frac{3Px^2}{lS_sh^2}$ $b = \frac{3Pl}{S_sh^2}$ Deflection at A: $f = \frac{3P}{bE}\left(\frac{l}{h}\right)^3$
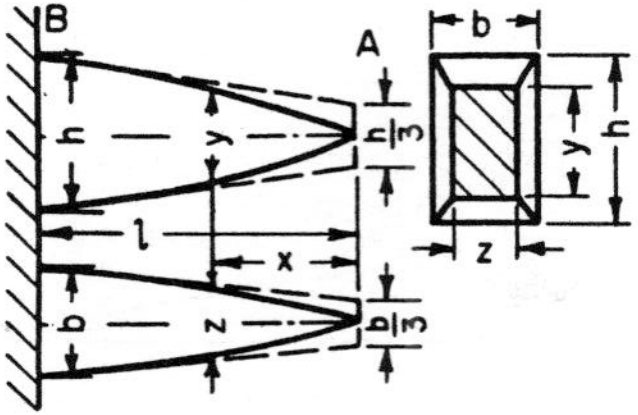	Rectangle: width (z) variable, depth (y) variable, $\frac{z}{y} = k$	Elevation: semicubic parabola Plan: semicubic parabola	$y^3 = \frac{3Px^2}{kS_sl}$ $z = ky$ $h = \sqrt[3]{\frac{3Pl}{kS_s}}$ $b = kh$

FIGURE 2.17 (*Continued*) Beams of uniform strength (in bending).

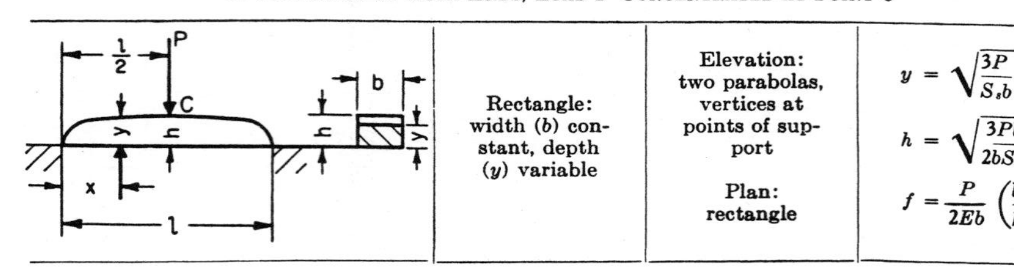

2. Fixed at One End, Load P Uniformly Distributed over l

Beam	Cross section	Elevation and plan	Formulas
	Circle: diam (y) variable	Elevation: semicubic parabola Plan: semicubic parabola	$y^3 = \frac{16P}{\pi l S_s} x^2$ $d = \sqrt[3]{\frac{16Pl}{\pi S_s}}$

3. Supported at Both Ends, Load P Concentrated at Point C

Beam	Cross section	Elevation and plan	Formulas
	Rectangle: width (b) constant, depth (y) variable	Elevation: two parabolas, vertices at points of support Plan: rectangle	$y = \sqrt{\frac{3P}{S_s b} x}$ $h = \sqrt{\frac{3Pl}{2bS_s}}$ $f = \frac{P}{2Eb} \left(\frac{l}{h}\right)^3$

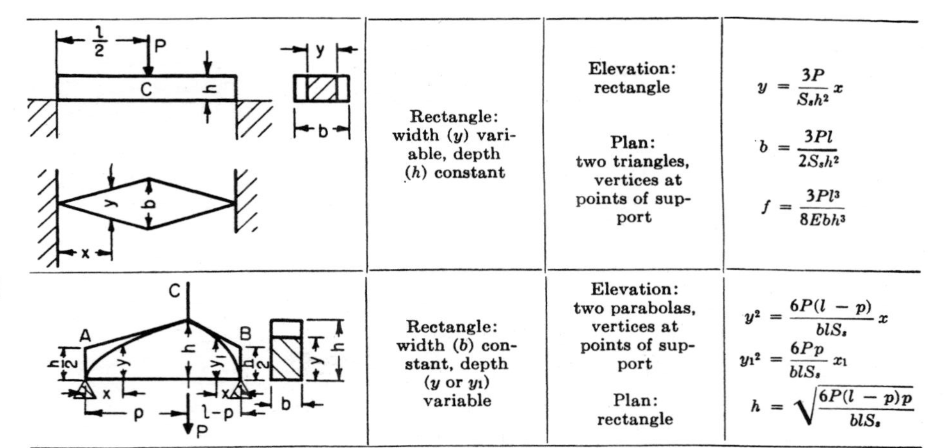

	Rectangle: width (y) variable, depth (h) constant	Elevation: rectangle Plan: two triangles, vertices at points of support	$y = \frac{3P}{S_s h^2} x$ $b = \frac{3Pl}{2S_s h^2}$ $f = \frac{3Pl^3}{8Ebh^3}$
	Rectangle: width (b) constant, depth (y or y_1) variable	Elevation: two parabolas, vertices at points of support Plan: rectangle	$y^2 = \frac{6P(l-p)}{blS_s} x$ $y_1^2 = \frac{6Pp}{blS_s} x_1$ $h = \sqrt{\frac{6P(l-p)p}{blS_s}}$

FIGURE 2.17 (*Continued*) Beams of uniform strength (in bending).

Beam	Cross section	Elevation and plan	Formulas
3. SUPPORTED AT BOTH ENDS, LOAD P MOVING ACROSS SPAN			
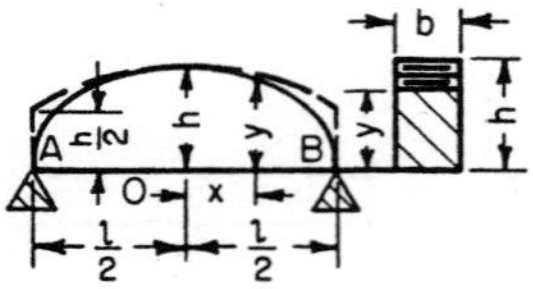 b, h, A, $\frac{h}{2}$, h, y, B, y, O, x, $\frac{l}{2}$, $\frac{l}{2}$	Rectangle: width (b) constant, depth (y) variable	Elevation: ellipse Major axis $= l$ Minor axis $= 2h$ Plan: rectangle	$\frac{x^2}{\left(\frac{l}{2}\right)^2} + \frac{y^2}{\frac{3Pl}{2bS_s}} = 1$ $h = \sqrt{\frac{3Pl}{2bS_s}}$

4. SUPPORTED AT BOTH ENDS, LOAD P UNIFORMLY DISTRIBUTED OVER l

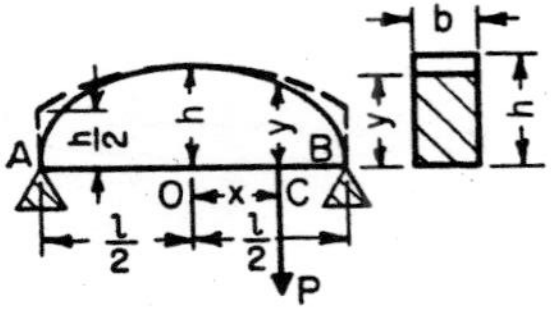 A, B, C, O, x, $\frac{l}{2}$, $\frac{l}{2}$, P, h, $\frac{h}{2}$, y, b	Rectangle: width (b) constant, depth (y) variable	Elevation: ellipse Plan: rectangle	$\frac{x^2}{\left(\frac{l}{2}\right)^2} + \frac{y^2}{\frac{3Pl}{4bS_s}} = 1$ $h = \sqrt{\frac{3Pl}{4bS_s}}$ Deflection at O: $f = \frac{1}{64}\frac{Pl^3}{EI}$ $= \frac{3}{16}\frac{P}{bE}\left(\frac{l}{h}\right)^3$

FIGURE 2.17 (*Continued*) Beams of uniform strength (in bending).

Beam	Cross section	Elevation and plan	Formulas
4. SUPPORTED AT BOTH ENDS, LOAD P UNIFORMLY DISTRIBUTED OVER l			
	Rectangle: width (y) variable, depth (h) constant	Elevation: rectangle Plan: two parabolas with vertices at center of span	$y = \frac{3P}{S_s h^2}\left(x - \frac{x^2}{l}\right)$ $b = \frac{3Pl}{4S_s h^2}$

FIGURE 2.17 (*Continued*) Beams of uniform strength (in bending).

ROLLING AND MOVING LOADS

Rolling and moving loads are loads that may change their position on a beam or beams. Figure 2.18 shows a beam with two equal concentrated moving loads, such as two wheels on a crane girder, or the wheels of a truck on a bridge. Because the maximum moment occurs where the shear is zero, the shear diagram shows that the maximum moment occurs under a wheel. Thus, with $x < a/2$:

$$R_1 = P\left(1 - \frac{2x}{l} + \frac{a}{l}\right)$$

$$M_2 = \frac{Pl}{2}\left(1 - \frac{a}{l} + \frac{2x}{l}\frac{a}{l} - \frac{4x^2}{l^2}\right)$$

$$R_2 = P\left(1 + \frac{2x}{l} - \frac{a}{l}\right)$$

$$M_1 = \frac{Pl}{2}\left(1 - \frac{a}{l} - \frac{2a^2}{l^2} + \frac{2x}{l}\frac{3a}{l} - \frac{4x^2}{l^2}\right)$$

$$M_2 \text{ max when } x = \tfrac{1}{4}a$$

$$M_1 \text{ max when } x = \tfrac{3}{4}a$$

$$M_{\max} = \frac{Pl}{2}\left(1 - \frac{a}{2l}\right)^2 = \frac{P}{2l}\left(l - \frac{a}{2}\right)^2$$

Figure 2.19 shows the condition when two equal loads are equally distant on opposite sides of the center of the beam. The moment is then equal under the two loads.

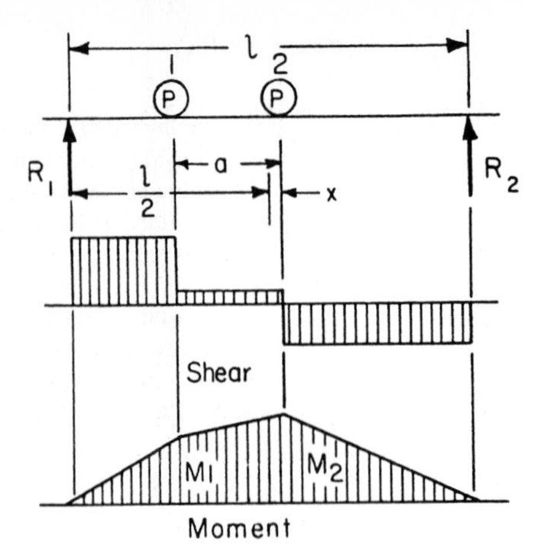

FIGURE 2.18 Two equal concentrated moving loads.

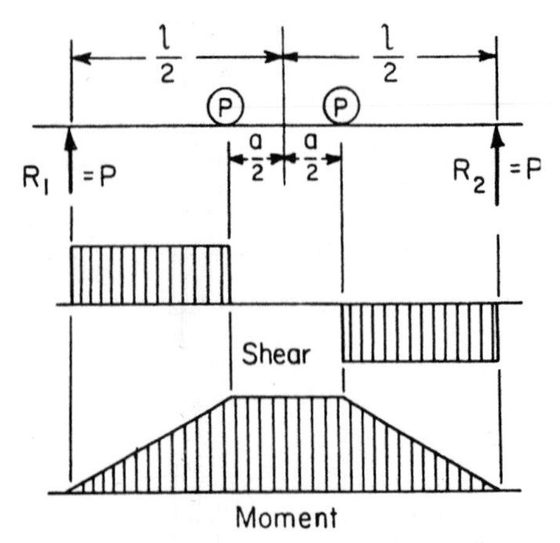

FIGURE 2.19 Two equal moving loads equally distant on opposite sides of the center.

If two moving loads are of unequal weight, the condition for maximum moment is the maximum moment occurring under the heavier wheel when the center of the beam bisects the distance between the resultant of the loads and the heavier wheel. Figure 2.20 shows this position and the shear and moment diagrams.

When several wheel loads constituting a system are on a beam or beams, the several wheels must be examined in turn to determine which causes the greatest moment. The position for the greatest moment that can occur under a given

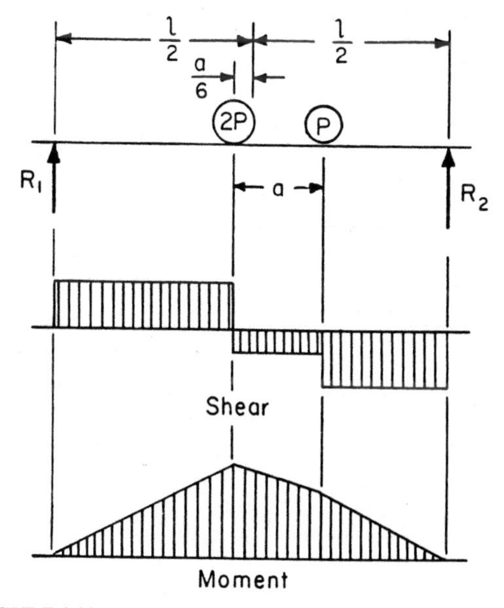

FIGURE 2.20 Two moving loads of unequal weight.

wheel is, as stated earlier, when the center of the span bisects the distance between the wheel in question and the resultant of all loads then on the span. The position for maximum shear at the support is when one wheel is passing off the span.

CURVED BEAMS

The application of the flexure formula for a straight beam to the case of a curved beam results in error. When all "fibers" of a member have the same center of curvature, the *concentric* or common type of curved beam exists (Fig. 2.21). Such a beam is defined by the Winkler–Bach theory. The stress at a point y units from the centroidal axis is

$$S = \frac{M}{AR}\left[1 + \frac{y}{Z(R + y)}\right]$$

M is the bending moment, positive when it increases curvature; y is positive when measured toward the convex side; A is the cross-sectional area; R is the radius of the centroidal axis; *Z is a cross-section property* defined by

$$Z = -\frac{1}{A}\int \frac{y}{R + y}\, dA$$

Analytical expressions for Z of certain sections are given in Table 2.4. Z can also be found by *graphical* integration methods (see any advanced strength book). The *neutral surface* shifts toward the center of curvature, or inside fiber, an amount equal to $e = ZR/(Z + 1)$. The Winkler–Bach

TABLE 2.4 Analytical Expressions for Z

Section	Expression
h, C, R	$Z = -1 + \frac{R}{h}\left(\ln \frac{R + C}{R - C}\right)$
r, R	$Z = -1 + 2\left(\frac{R}{r}\right)\left[\frac{R}{r} - \sqrt{\left(\frac{R}{r}\right)^2 - 1}\right]$
t, b, C_1, C_3, C_2, R	$Z = -1 + \frac{R}{A}\left[t \ln (R + C_1) + (b - t) \ln (R - C_0) - b \ln (R - C_2)\right]$ and $A = tC_1 - (b - t)\, C_3 + bC_2$
t, b, C_1, C_2, R	$Z = -1 + \frac{R}{A}\left[b \ln \left(\frac{R + C_2}{R - C_2}\right) + (t - b) \ln \left(\frac{R + C_1}{R - C_1}\right)\right]$ $A = 2\,[(t - b)C_1 + bC_2]$

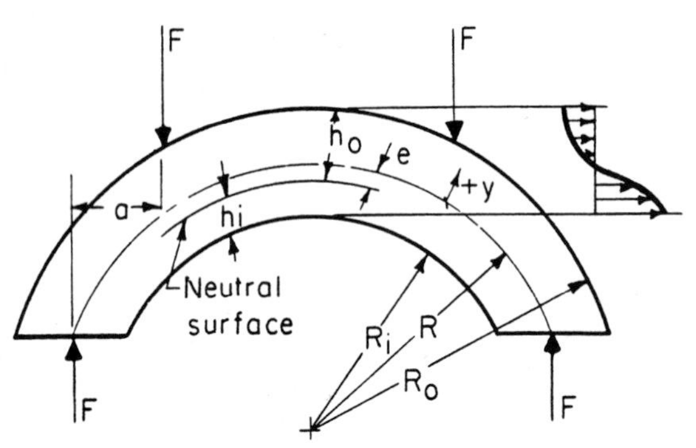

FIGURE 2.21 Curved beam.

theory, though practically satisfactory, disregards radial stresses as well as lateral deformations and assumes pure bending. The *maximum stress* occurring on the inside fiber is $S = Mh_i/AeR_i$, whereas that on the outside fiber is $S = Mh_0/AeR_0$.

The *deflection* in curved beams can be computed by means of the moment-area theory.

The resultant deflection is then equal to $\Delta_0 = \sqrt{\Delta_x^2 + \Delta_y^2}$ in the direction defined by $\tan \theta = \Delta_y / \Delta_x$. Deflections can also be found conveniently by use of *Castigliano's theorem.* It states that in an elastic system the displacement in the direction of a force (or couple) and due to that force (or couple) is the partial derivative of the strain energy with respect to the force (or couple).

A quadrant of radius R is fixed at one end as shown in Fig. 2.22. The force F is applied in the radial direction at free-end B. Then, the deflection of B is

By moment area,

$$y = R\sin\theta \quad x = R(1 - \cos\theta)$$

$$ds = Rd\theta \quad M = FR\sin\theta$$

$$_B\Delta_x = \frac{\pi FR^3}{4EI} \qquad _B\Delta_y = -\frac{FR^3}{2EI}$$

$$\text{and} \quad \Delta_B = \frac{FR^3}{2EI}\sqrt{1 + \frac{\pi^2}{4}}$$

$$\text{at} \qquad \theta_x = \tan^{-1}\left(-\frac{FR^3}{2EI} \times \frac{4EI}{\pi FR^3}\right)$$

$$= \tan^{-1}\frac{2}{\pi}$$

$$= 32.5°$$

By Castigliano,

$$_B\Delta_x = \frac{\pi FR^3}{4EI} \qquad _B\Delta_y = -\frac{FR^3}{2EI}$$

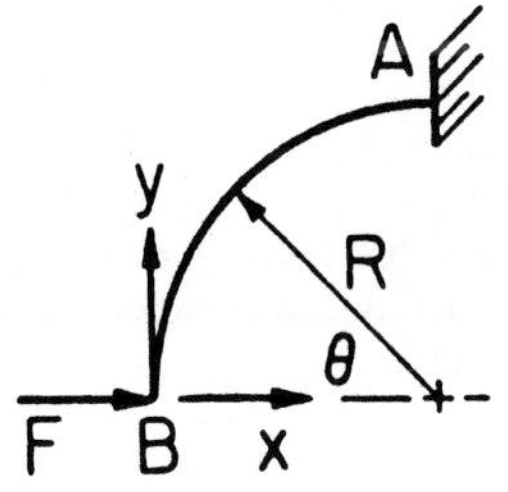

FIGURE 2.22 Quadrant with fixed end.

Eccentrically Curved Beams

These beams (Fig. 2.23) are bounded by arcs having different centers of curvature. In addition, it is possible for either radius to be the larger one. The one in which the section depth shortens as the central section is approached may be called the *arch beam*. When the central section is the largest, the beam is of the crescent type.

Crescent I denotes the beam of larger outside radius and *crescent II* of larger inside radius. The stress at the *central section* of such beams may be found from $S = KMC/I$. In the case of rectangular cross section, the equation becomes $S = 6KM/bh^2$, where M is the bending moment, b is the width of the beam section, and h its height. The *stress factors*, K for the *inner boundary*, established from photoelastic data, are given in Table 2.5. The outside radius

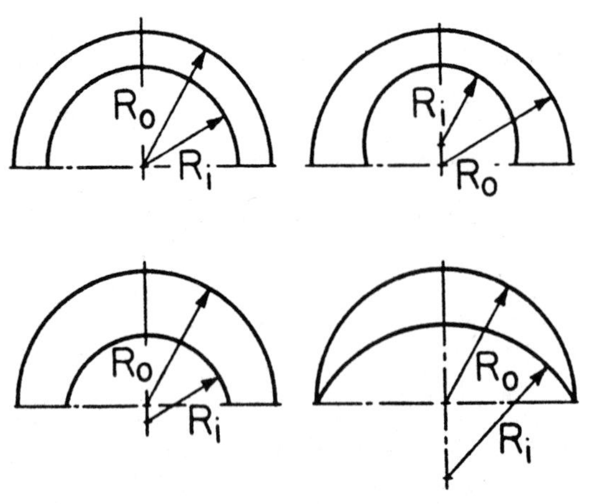

FIGURE 2.23 Eccentrically curved beams.

TABLE 2.5 Stress Factors for Inner Boundary at Central Section *(see Fig. 2.23)*

1. For the arch-type beams

(a) $K = 0.834 + 1.504 \dfrac{h}{R_o + R_i}$ if $\dfrac{R_o + R_i}{h} < 5$

(b) $K = 0.899 + 1.181 \dfrac{h}{R_o + R_i}$ if $5 < \dfrac{R_o + R_i}{h} < 10$

(c) In the case of larger section ratios use the equivalent beam solution

2. For the crescent I-type beams

(a) $K = 0.570 + 1.536 \dfrac{h}{R_o + R_i}$ if $\dfrac{R_o + R_i}{h} < 2$

(b) $K = 0.959 + 0.769 \dfrac{h}{R_o + R_i}$ if $2 < \dfrac{R_o + R_i}{h} < 20$

(c) $K = 1.092 \left(\dfrac{h}{R_o + R_i}\right)^{0.0298}$ if $\dfrac{R_o + R_i}{h} > 20$

3. For the crescent II-type beams

(a) $K = 0.897 + 1.098 \dfrac{h}{R_o + R_i}$ if $\dfrac{R_o + R_i}{h} < 8$

(b) $K = 1.119 \left(\dfrac{h}{R_o + R_i}\right)^{0.0378}$ if $8 < \dfrac{R_o + R_i}{h} < 20$

(c) $K = 1.081 \left(\dfrac{h}{R_o + R_i}\right)^{0.0270}$ if $\dfrac{R_o + R_i}{h} > 20$

is denoted by R_o and the inside by R_i. The geometry of crescent beams is such that the stress can be larger in *off-center sections*. The stress at the central section determined above must then be multiplied by the *position factor k*, given in Table 2.6. As in the concentric beam, the *neutral surface* shifts slightly toward the inner boundary. (See Vidosic, "Curved Beams with Eccentric Boundaries," *Transactions of the ASME*, *79*, pp. 1317–1321.)

ELASTIC LATERAL BUCKLING OF BEAMS

When lateral buckling of a beam occurs, the beam undergoes a combination of twist and out-of-plane bending (Fig. 2.24). For a simply supported beam of rectangular cross section subjected to uniform bending, buckling occurs at the critical bending moment, given by

$$M_{cr} = \frac{\pi}{L}\sqrt{EI_yGJ}$$

where L = unbraced length of the member

E = modulus of elasticity

I_y = moment of inertial about minor axis

G = shear modulus of elasticity

J = torsional constant

The critical moment is proportional to both the lateral bending stiffness EI_y/L and the torsional stiffness of the member GJ/L.

For the case of an open section, such as a wide-flange or I-beam section, warping rigidity can provide additional torsional stiffness. Buckling of a simply supported beam of open cross section subjected to uniform bending occurs at the critical bending moment, given by

$$M_{cr} = \frac{\pi}{L}\sqrt{EI_y\left(GJ + EC_w\frac{\pi^2}{L^2}\right)}$$

where C_w is the warping constant, a function of cross-sectional shape and dimensions (Fig. 2.25).

In the preceding equations, the distribution of bending moment is assumed to be uniform. For the case of a nonuniform bending-moment gradient, buckling often occurs at

TABLE 2.6 Crescent-Beam Position Stress Factors *(see Fig. 2.23)*†

Angle θ, degree	k Inner	k Outer
10	$1 + 0.055\, H/h$	$1 + 0.03\, H/h$
20	$1 + 0.164\, H/h$	$1 + 0.10\, H/h$
30	$1 + 0.365\, H/h$	$1 + 0.25\, H/h$
40	$1 + 0.567\, H/h$	$1 + 0.467\, H/h$
50	$1.521 - \dfrac{(0.5171 - 1.382\, H/h)^{1/2}}{1.382}$	$1 + 0.733\, H/h$
60	$1.756 - \dfrac{(0.2416 - 0.6506\, H/h)^{1/2}}{0.6506}$	$1 + 1.123\, H/h$
70	$2.070 - \dfrac{(0.4817 - 1.298\, H/h)^{1/2}}{0.6492}$	$1 + 1.70\, H/h$
80	$2.531 - \dfrac{(0.2939 - 0.7084\, H/h)^{1/2}}{0.3542}$	$1 + 2.383\, H/h$
90		$1 + 3.933\, H/h$

†Note: All formulas are valid for $0 < H/h \leq 0.325$. Formulas for the inner boundary, except for 40 degrees, may be used to $H/h \leq 0.36$. H = distance between centers.

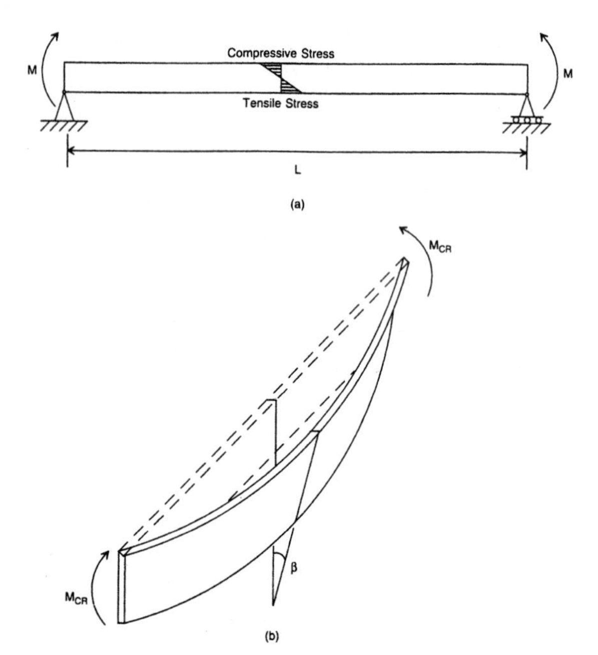

FIGURE 2.24 (*a*) Simple beam subjected to equal end moments. (*b*) Elastic lateral buckling of the beam.

a larger critical moment. Approximation of this critical bending moment, M_{cr}' may be obtained by multiplying M_{cr} given by the previous equations by an amplification factor

$$M_{cr}' = C_b M_{cr}$$

$$\text{where } C_b = \frac{12.5M_{\max}}{2.5M_{\max} + 3M_A + 4M_B + 3M_C}$$

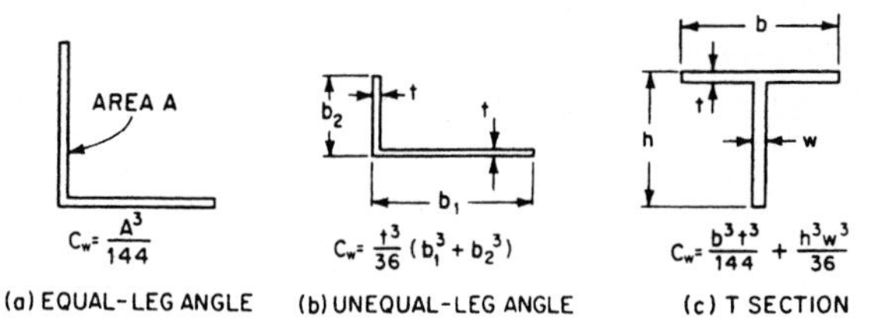

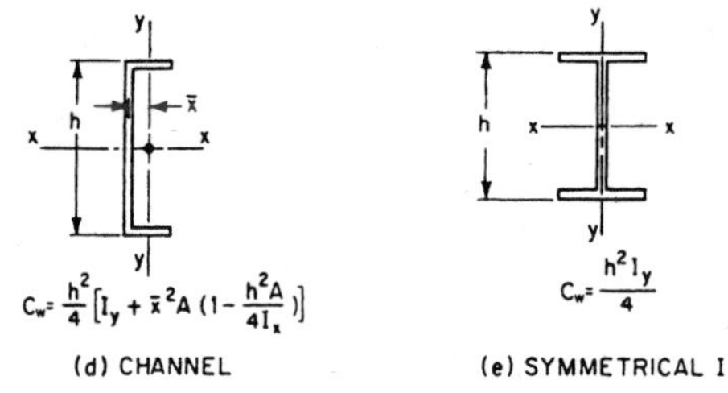

FIGURE 2.25 Torsion-bending constants for torsional buckling. A = cross-sectional area; I_x = moment of inertia about x–x axis; I_y = moment of inertia about y–y axis. (After McGraw-Hill, New York). Bleich, F., *Buckling Strength of Metal Structures*.

and M_{max} = absolute value of maximum moment in the unbraced beam segment

M_A = absolute value of moment at quarter point of the unbraced beam segment

M_B = absolute value of moment at centerline of the unbraced beam segment

M_C = absolute value of moment at three-quarter point of the unbraced beam segment

C_b equals 1.0 for unbraced cantilevers and for members where the moment within a significant portion of the unbraced segment is greater than, or equal to, the larger of the segment end moments.

COMBINED AXIAL AND BENDING LOADS

For short beams, subjected to both transverse and axial loads, the stresses are given by the principle of superposition if the deflection due to bending may be neglected without serious error. That is, the total stress is given with sufficient accuracy at any section by the sum of the axial stress and the bending stresses. The maximum stress, lb/in^2 (MPa), equals

$$f = \frac{P}{A} + \frac{Mc}{I}$$

where P = axial load, lb (N)

A = cross-sectional area, in^2 (mm^2)

M = maximum bending moment, in lb (Nm)

c = distance from neutral axis to outermost fiber at the section where maximum moment occurs, in (mm)

I = moment of inertia about neutral axis at that section, in^4 (mm^4)

When the deflection due to bending is large and the axial load produces bending stresses that cannot be neglected, the maximum stress is given by

$$f = \frac{P}{A} + (M + Pd)\frac{c}{I}$$

where d is the deflection of the beam. For axial compression, the moment Pd should be given the same sign as M; and for tension, the opposite sign, but the minimum value of $M + Pd$ is zero. The deflection d for axial compression and bending can be closely approximated by

$$d = \frac{d_o}{1 - (P/P_c)}$$

where d_o = deflection for the transverse loading alone, in (mm); and P_c = critical buckling load $\pi^2 EI / L^2$, lb (N).

UNSYMMETRICAL BENDING

When a beam is subjected to loads that do not lie in a plane containing a principal axis of each cross section, unsymmetrical bending occurs. Assuming that the bending axis of the beam lies in the plane of the loads, to preclude torsion, and that the loads are perpendicular to the bending axis, to preclude axial components, the stress, lb/in^2 (MPa), at any point in a cross section is

$$f = \frac{M_x y}{I_x} + \frac{M_y x}{I_y}$$

where M_x = bending moment about principal axis XX, in lb (Nm)

M_y = bending moment about principal axis YY, in lb (Nm)

x = distance from point where stress is to be computed to YY axis, in (mm)

y = distance from point to XX axis, in (mm)

I_x = moment of inertia of cross section about XX, in (mm^4)

I_y = moment of inertia about YY, in (mm^4)

If the plane of the loads makes an angle θ with a principal plane, the neutral surface forms an angle α with the other principal plane such that

$$\tan \alpha = \frac{I_x}{I_y} \tan \theta$$

ECCENTRIC LOADING

If an eccentric longitudinal load is applied to a bar in the plane of symmetry, it produces a bending moment Pe, where e is the distance, in (mm), of the load P from the centroidal axis. The total unit stress is the sum of this moment and the stress due to P applied as an axial load:

$$f = \frac{P}{A} \pm \frac{Pec}{I} = \frac{P}{A}\left(1 \pm \frac{ec}{r^2}\right)$$

where A = cross-sectional area, in^2 (mm^2)

c = distance from neutral axis to outermost fiber, in (mm)

I = moment of inertia of cross section about neutral axis, in^4 (mm^4)

r = radius of gyration = $\sqrt{I/A}$, in (mm)

Figure 2.1 gives values of the radius of gyration for several cross sections.

If there is to be no tension on the cross section under a compressive load, e should not exceed r^2/c. For a rectangular section with width b, and depth d, the eccentricity, therefore, should be less than $b/6$ and $d/6$ (i.e., the load should not be applied outside the middle third). For a circular cross section with diameter D, the eccentricity should not exceed $D/8$.

When the eccentric longitudinal load produces a deflection too large to be neglected in computing the bending stress, account must be taken of the additional bending moment Pd, where d is the deflection, in (mm). This deflection may be closely approximated by

$$d = \frac{4eP/P_c}{\pi(1 - P/P_c)}$$

P_c is the critical buckling load π^2EI/L^2, lb (N).

If the load P, does not lie in a plane containing an axis of symmetry, it produces bending about the two principal axes through the centroid of the section. The stresses, lb/in^2 (MPa), are given by

$$f = \frac{P}{A} + \frac{Pe_xc_x}{I_y} + \frac{Pe_yc_y}{I_x}$$

where A = cross-sectional area, in^2 (mm^2)

e_x = eccentricity with respect to principal axis YY, in (mm)

e_y = eccentricity with respect to principal axis XX, in (mm)

c_x = distance from YY to outermost fiber, in (mm)

c_y = distance from XX to outermost fiber, in (mm)

I_x = moment of inertia about XX, in^4 (mm^4)

I_y = moment of inertia about YY, in^4 (mm^4)

The principal axes are the two perpendicular axes through the centroid for which the moments of inertia are a maximum or a minimum and for which the products of inertia are zero.

NATURAL CIRCULAR FREQUENCIES AND NATURAL PERIODS OF VIBRATION OF PRISMATIC BEAMS

Figure 2.26 shows the characteristic shape and gives constants for determination of natural circular frequency ω and natural period T, for the first four modes of cantilever, simply supported, fixed-end, and fixed-hinged beams. To obtain ω, select the appropriate constant from Fig. 2.26 and multiply it by $\sqrt{EI/wL^4}$. To get T, divide the appropriate constant by $\sqrt{EI/wL^4}$.

In these equations,

ω = natural frequency, rad/s

W = beam weight, lb per linear ft (kg per linear m)

L = beam length, ft (m)

TYPE OF SUPPORT	FUNDAMENTAL MODE	SECOND MODE	THIRD MODE	FOURTH MODE
CANTILEVER	L	0.774L	0.5L, 0.132L	0.356L, 0.094L, 0.644L
$\omega\sqrt{wL^4/EI}$ =	20.0	125	350	684
$T\sqrt{EI/wL^4}$ =	0.315	0.0503	0.0180	0.0092
SIMPLE	L	0.5L	$\frac{L}{3}$, $\frac{L}{3}$	$\frac{L}{4}$, $\frac{L}{4}$, $\frac{L}{2}$
$\omega\sqrt{wL^4/EI}$ =	56.0	224	502	897
$T\sqrt{EI/wL^4}$ =	0.112	0.0281	0.0125	0.0070
FIXED	L	$\frac{L}{2}$	0.359L, 0.359L	0.278L, $\frac{L}{2}$, 0.278L
$\omega\sqrt{wL^4/EI}$ =	127	350	684	1,133
$T\sqrt{EI/wL^4}$ =	0.0496	0.0180	0.0092	0.0056
FIXED - HINGED	L	0.56L	0.384L, 0.308L	0.294L, 0.235L, 0.529L
$\omega\sqrt{wL^4/EI}$ =	87.2	283	591	1,111
$T\sqrt{EI/wL^4}$ =	0.0722	0.0222	0.0106	0.0062

FIGURE 2.26 Coefficients for computing natural circular frequencies and natural periods of vibration of prismatic beams.

E = modulus of elasticity, lb/in^2 (MPa)

I = moment of inertia of beam cross section, in^4 (mm^4)

T = natural period, s

To determine the characteristic shapes and natural periods for beams with variable cross section and mass, use the Rayleigh method. Convert the beam into a lumped-mass system by dividing the span into elements and assuming the mass of each element to be concentrated at its center. Also, compute all quantities, such as deflection and bending moment, at the center of each element. Start with an assumed characteristic shape.

CHAPTER 3

COLUMN FORMULAS

GENERAL CONSIDERATIONS

Columns are structural members subjected to direct compression. All columns can be grouped into the following three classes:

1. *Compression blocks* are so short (with a slenderness ratio—that is, unsupported length divided by the least radius of gyration of the member—below 30) that bending is not potentially occurring.
2. Columns so slender that bending under load is a given are termed *long columns* and are defined by Euler's theory.
3. Intermediate-length columns, often used in structural practice, are called *short columns*.

Long and short columns usually fail by buckling when their *critical load* is reached. Long columns are analyzed using Euler's column formula, namely,

$$P_{\text{cr}} = \frac{n\pi^2 EI}{l^2} = \frac{n\pi^2 EA}{(l/r)^2}$$

In this formula, the coefficient n accounts for end conditions. When the column is pivoted at both ends, $n = 1$; when one end is fixed and the other end is rounded, $n = 2$; when both ends are fixed, $n = 4$; and when one end is fixed and the other is free, $n = 0.25$. The slenderness ratio separating long columns from short columns depends on the modulus of elasticity and the yield strength of the column material. When Euler's formula results in $(P_{\text{cr}}/A) > S_y$, strength instead of buckling causes failure, and the column ceases to be long. In quick estimating numbers, this *critical slenderness ratio* falls between 120 and 150. Table 3.1 gives additional column data based on Euler's formula.

TABLE 3.1 Strength of Round-Ended Columns According to Euler's Formula*

Material†	Cast iron	Wrought iron	Low-carbon steel	Medium-carbon steel
Ultimate compressive strength, lb/in^2	107,000	53,400	62,600	89,000
Allowable compressive stress, lb/in^2 (maximum)	7,100	15,400	17,000	20,000
Modulus of elasticity	14,200,000	28,400,000	30,600,000	31,300,000
Factor of safety	8	5	5	5
Smallest I allowable at worst section, in^4	$\frac{Pl^2}{17,500,000}$	$\frac{Pl^2}{56,000,000}$	$\frac{Pl^2}{60,300,000}$	$\frac{Pl^2}{61,700,000}$
Limit of ratio, $l/r >$	50.0	60.6	59.4	55.6
Rectangle $\left(r = b\sqrt{1/12}\right)$, $l/b >$	14.4	17.5	17.2	16.0
Circle $\left(r = \frac{1}{4}d\right)$, $l/d >$	12.5	15.2	14.9	13.9
Circular ring of small thickness $\left(r = d\sqrt{1/8}\right)$, $l/d >$	17.6	21.4	21.1	19.7

* (P = allowable load, lb; l = length of column, in; b = smallest dimension of a rectangular section, in; d = diameter of a circular section, in; r = least radius of gyration of section.)

† To convert to SI units, use: $1b/in^2 \times 6.894$ = kPa; $in^4 \times (25.4)^4 = mm^4$.

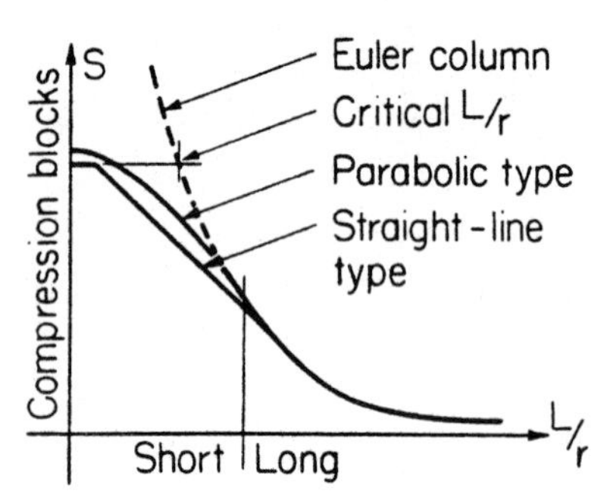

FIGURE 3.1 *L/r* plot for columns.

SHORT COLUMNS

Stress in short columns can be considered to be partly due to compression and partly due to bending. Empirical, rational expressions for column stress are, in general, based on the assumption that the permissible stress must be reduced below that which could be permitted were it due to compression only. The manner in which this reduction is made determines the type of equation and the slenderness ratio beyond which the equation does not apply. Figure 3.1 shows the curves for this situation. Typical column formulas are given in Table 3.2.

ECCENTRIC LOADS ON COLUMNS

When short blocks are loaded eccentrically in compression or in tension, that is, not through the center of gravity (cg), a combination of axial and bending stress results. The maximum unit stress S_M is the algebraic sum of these two unit stresses.

TABLE 3.2 Typical Short-Column Formulas

Formula	Material	Code	Slenderness ratio
$S_w = 17{,}000 - 0.485\left(\frac{l}{r}\right)^2$	Carbon steels	AISC	$\frac{l}{r} < 120$
$S_w = 16{,}000 - 70\left(\frac{l}{r}\right)$	Carbon steels	Chicago	$\frac{l}{r} < 120$
$S_w = 15{,}000 - 50\left(\frac{l}{r}\right)$	Carbon steels	AREA	$\frac{l}{r} < 150$
$S_w = 19{,}000 - 100\left(\frac{l}{r}\right)$	Carbon steels	Am. Br. Co.	$60 < \frac{l}{r} < 120$
$^{\dagger}S_{cr} = 135{,}000 - \frac{15.9}{c}\left(\frac{l}{r}\right)^2$	Alloy-steel tubing	ANC	$\frac{l}{\sqrt{c}r} < 65$
$S_w = 9{,}000 - 40\left(\frac{l}{r}\right)$	Cast iron	NYC	$\frac{l}{r} < 70$

TABLE 3.2 Typical Short-Column Formulas (*Continued*)

Formula	Material	Code	Slenderness ratio
$^{\dagger}S_{cr} = 34{,}500 - \frac{245}{\sqrt{c}}\left(\frac{l}{r}\right)$	2017ST aluminum	ANC	$\frac{1}{\sqrt{c}r} < 94$
$^{\dagger}S_{cr} = 5{,}000 - \frac{0.5}{c}\left(\frac{l}{r}\right)^2$	Spruce	ANC	$\frac{1}{\sqrt{c}r} < 72$
$^{\dagger}S_{cr} = S_y\left[1 - \frac{S_y}{4n\pi^2 E}\left(\frac{l}{r}\right)^2\right]$	Steels	Johnson	$\frac{l}{r} < \sqrt{\frac{2n\pi^2 E}{S_y}}$
$^{\ddagger}S_{cr} = \frac{S_y}{1 + \frac{ec}{r^2}\sec\left(\frac{l}{r}\sqrt{\frac{P}{4AE}}\right)}$	Steels	Secant	$\frac{l}{r} <$ critical

† S_{cr} = theoretical maximum; c = end fixity coefficient; c = 2, both ends pivoted; c = 2.86, one pivoted, other fixed; c = 4, both ends fixed; c = 1 one fixed, one free.
‡ is initial eccentricity at which load is applied to center of column cross section.

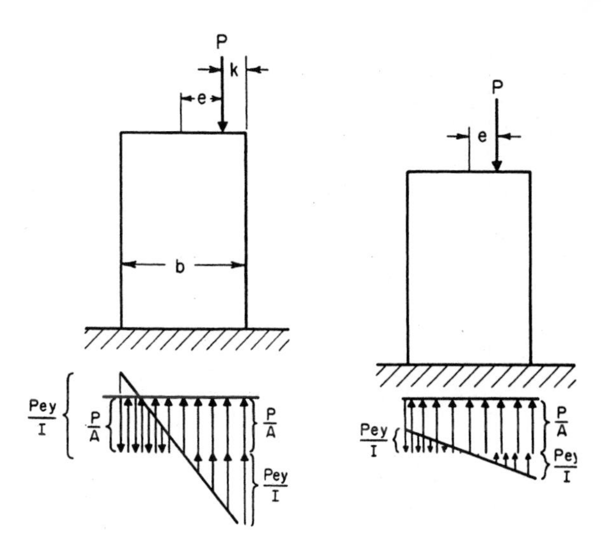

FIGURE 3.2 Load plot for columns.

In Fig. 3.2, a load, P, acts in a line of symmetry at the distance e from cg; r = radius of gyration. The unit stresses are (1) S_c, due to P, as if it acted through cg, and (2) S_b, due to the bending moment of P acting with a leverage of e about cg. Thus, unit stress, S, at any point y is

$$S = S_c \pm S_b$$

$$= (P/A) \pm Pey/I$$

$$= S_c(1 \pm ey/r^2)$$

y is positive for points on the same side of cg as P, and negative on the opposite side. For a *rectangular cross section* of width b, the maximum stress, $S_M = S_c(1 + 6e/b)$. When P is outside the middle third of width b and is a compressive load, tensile stresses occur.

For a *circular cross section* of diameter d, $S_M = S_c(1 + 8e/d)$. The stress due to the weight of the solid modifies these relations.

Note that in these formulas e is measured from the gravity axis and gives tension when e is greater than one-sixth the width (measured in the same direction as e), for rectangular sections, and when greater than one-eighth the diameter, for solid circular sections.

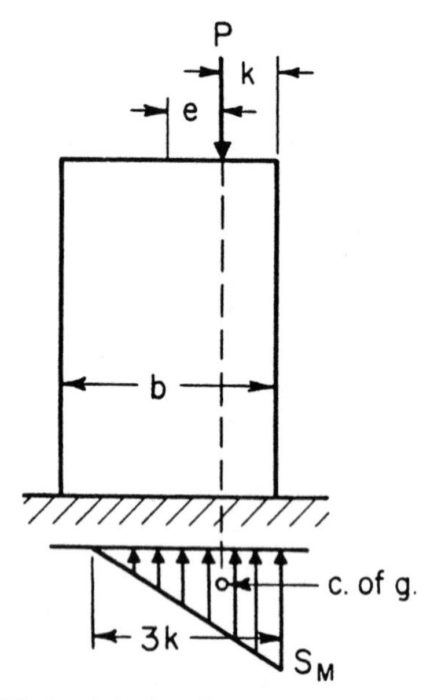

FIGURE 3.3 Load plot for columns.

If, as in certain classes of masonry construction, the *material cannot withstand tensile stress* and, thus, no tension can occur, the center of moments (Fig. 3.3) is taken at the center of stress. For a *rectangular section*, P acts at distance k from the nearest edge. Length under compression = $3k$, and $S_M = \frac{2}{3}P/hk$. For a *circular section*, $S_M = [0.372 + 0.056(k/r)]P/k\sqrt{rk}$, where r = radius and k = distance of P from circumference. For a *circular ring*, S = average compressive stress on cross section produced by P; e = eccentricity of P; z = length of diameter under compression (Fig. 3.4). Values of z/r and of the ratio of S_{max} to average S are given in Tables 3.3 and 3.4.

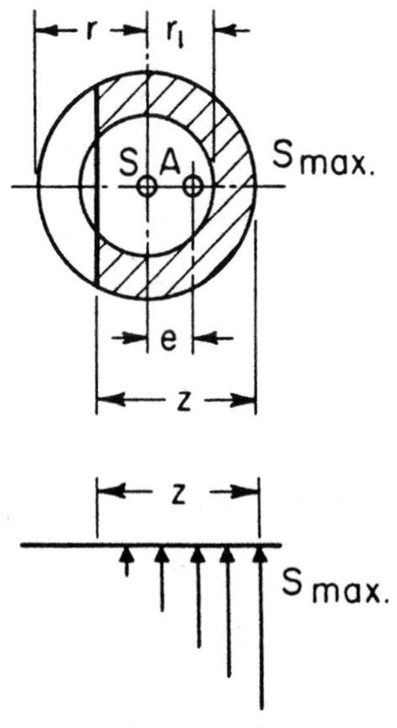

FIGURE 3.4 Circular column load plot.

TABLE 3.3 Values of the Ratio z/r

(See Fig. 3.5)

$\frac{e}{r}$	$\frac{r_1}{r}$ 0.0	0.5	0.6	0.7	0.8	0.9	1.0	$\frac{e}{r}$
0.25	2.00							0.25
0.30	1.82							0.30
0.35	1.66	1.89	1.98					0.35
0.40	1.51	1.75	1.84	1.93				0.40
0.45	1.37	1.61	1.71	1.81	1.90			0.45
0.50	1.23	1.46	1.56	1.66	1.78	1.89	2.00	0.50
0.55	1.10	1.29	1.39	1.50	1.62	1.74	1.87	0.55
0.60	0.97	1.12	1.21	1.32	1.45	1.58	1.71	0.60
0.65	0.84	0.94	1.02	1.13	1.25	1.40	1.54	0.65
0.70	0.72	0.75	0.82	0.93	1.05	1.20	1.35	0.70
0.75	0.59	0.60	0.64	0.72	0.85	0.99	1.15	0.75
0.80	0.47	0.47	0.48	0.52	0.61	0.77	0.94	0.80
0.85	0.35	0.35	0.35	0.36	0.42	0.55	0.72	0.85
0.90	0.24	0.24	0.24	0.24	0.24	0.32	0.49	0.90
0.95	0.12	0.12	0.12	0.12	0.12	0.12	0.25	0.95

TABLE 3.4 Values of the Ratio S_{max}/S_{avg}

(In determining S average, use load P divided by total area of cross section)

	$\frac{r_1}{r}$							
$\frac{e}{r}$	0.0	0.5	0.6	0.7	0.8	0.9	1.0	$\frac{e}{r}$
0.00	1.00	1.00	1.00	1.00	1.00	1.00	1.00	0.00
0.05	1.20	1.16	1.15	1.13	1.12	1.11	1.10	0.05
0.10	1.40	1.32	1.29	1.27	1.24	1.22	1.20	0.10
0.15	1.60	1.48	1.44	1.40	1.37	1.33	1.30	0.15
0.20	1.80	1.64	1.59	1.54	1.49	1.44	1.40	0.20
0.25	2.00	1.80	1.73	1.67	1.61	1.55	1.50	0.25
0.30	2.23	1.96	1.88	1.81	1.73	1.66	1.60	0.30
0.35	2.48	2.12	2.04	1.94	1.85	1.77	1.70	0.35
0.40	2.76	2.29	2.20	2.07	1.98	1.88	1.80	0.40
0.45	3.11	2.51	2.39	2.23	2.10	1.99	1.90	0.45

The *kern* is the area around the center of gravity of a cross section within which any load applied produces stress of only one sign throughout the entire cross section. Outside the kern, a load produces stresses of different sign. Figure 3.5 shows kerns (shaded) for various sections.

For a *circular ring*, the radius of the kern $r = D[1+(d/D)^2]/8$.

For a *hollow square* (H and h = lengths of outer and inner sides), the kern is a square similar to Fig. 3.5*a*, where

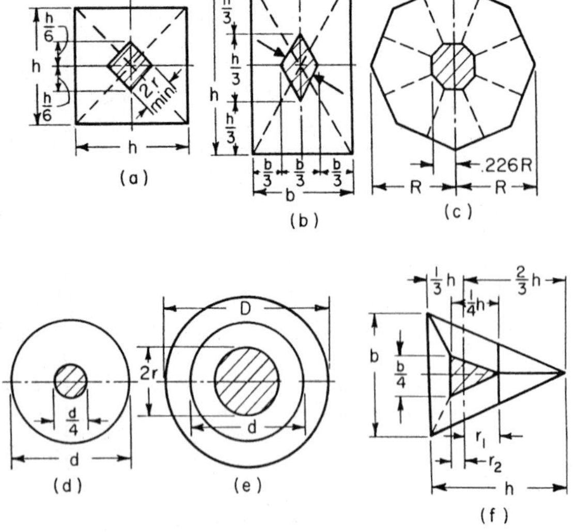

FIGURE 3.5 Column characteristics.

$$r_{\min} = \frac{H}{6}\frac{1}{\sqrt{2}}\left[1 + \left(\frac{h}{H}\right)^2\right] = 0.1179H\left[1 = \left(\frac{h}{H}\right)^2\right]$$

For a *hollow octagon*, R_a and R_i = radii of circles circumscribing the outer and inner sides; thickness of wall = $0.9239(R_a - R_i)$; and the kern is an octagon similar to Fig. 3.5*c*, where $0.2256R$ becomes $0.2256R_a[1 + (R_i/R_a)^2]$.

COLUMN BASE PLATE DESIGN

Base plates are usually used to distribute column loads over a large enough area of supporting concrete construction that the design bearing strength of the concrete is not exceeded. The factored load, P_u, is considered to be uniformly distributed under a base plate.

The nominal bearing strength f_p kip/in² or ksi (MPa) of the concrete is given by

$$f_p = 0.85f'_c\sqrt{\frac{A_1}{A_1}} \quad \text{and} \quad \sqrt{\frac{A_2}{A_1}} \leq 2$$

where f'_c = specified compressive strength of concrete, ksi (MPa)

A_1 = area of the base plate, in² (mm²)

A_2 = area of the supporting concrete that is geometrically similar to and concentric with the loaded area, in² (mm²)

In most cases, the bearing strength, f_p is $0.85f'_c$, when the concrete support is slightly larger than the base plate or $1.7f'_c$, when the support is a spread footing, pile cap, or mat foundation. Therefore, the required area of a base plate for a factored load P_u is

$$A_1 = \frac{P_u}{\phi_c 0.85 f'_c}$$

where ϕ_c is the strength reduction factor = 0.6. For a wide-flange column, A_1 should not be less than $b_f d$, where b_f is the flange width, in (mm), and d is the depth of column, in (mm).

The length N, in (mm), of a rectangular base plate for a wide-flange column may be taken in the direction of d as

$$N = \sqrt{A_1} + \Delta > d \qquad \text{or} \qquad \Delta = 0.5(0.95d - 0.80b_f)$$

The width B, in (mm), parallel to the flanges, then, is

$$B = \frac{A_1}{N}$$

The thickness of the base plate t_p, in (mm), is the largest of the values given by the equations that follow:

$$t_p = m\sqrt{\frac{2P_u}{0.9F_y BN}}$$

$$t_p = n\sqrt{\frac{2P_u}{0.9F_y BN}}$$

$$t_p = \lambda n'\sqrt{\frac{2P_u}{0.9F_y BN}}$$

where m = projection of base plate beyond the flange and parallel to the web, in (mm)

$= (N - 0.95d)/2$

n = projection of base plate beyond the edges of the flange and perpendicular to the web, in (mm)

$= (B - 0.80b_f)/2$

$n' = \sqrt{(db_f)}/4$

$\lambda = (2\sqrt{X})/[1 + \sqrt{(1 - X)}] \leq 1.0$

$X = [(4\, db_f)/(d + b_f)^2][Pu/(\phi \times 0.85 f'_c A_1)]$

AMERICAN INSTITUTE OF STEEL CONSTRUCTION ALLOWABLE-STRESS DESIGN APPROACH

The lowest columns of a structure usually are supported on a concrete foundation. The area, in square inches (square millimeters), required is found from:

$$A = \frac{P}{F_P}$$

where P is the load, kip (N) and F_p is the allowable bearing pressure on support, ksi (MPa).

The allowable pressure depends on strength of concrete in the foundation and relative sizes of base plate and concrete support area. If the base plate occupies the full area of the support, $F_p = 0.35f'_c$, where f'_c is the 28-day compressive strength of the concrete. If the base plate covers less than the full area, $F_P = 0.35f'_c \sqrt{A_2/A_1} \leq 0.70f'_c$, where A_1 is the base-plate area ($B \times N$), and A_2 is the full area of the concrete support.

Eccentricity of loading or presence of bending moment at the column base increases the pressure on some parts of the base plate and decreases it on other parts. To compute these effects, the base plate may be assumed completely rigid so that the pressure variation on the concrete is linear.

Plate thickness may be determined by treating projections m and n of the base plate beyond the column as cantilevers.

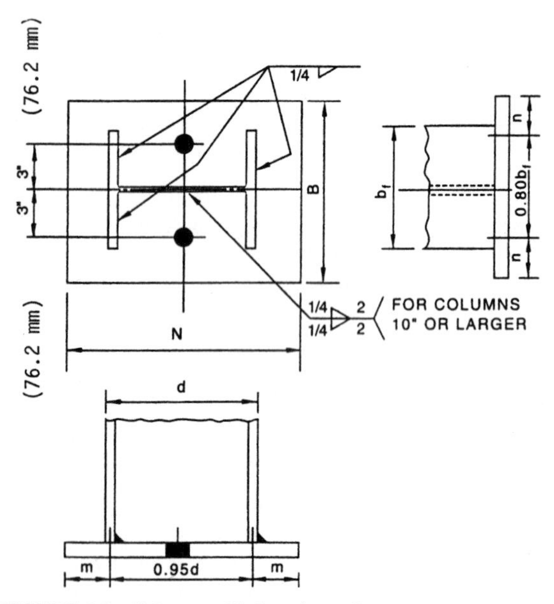

FIGURE 3.6 Column welded to a base plate.

The cantilever dimensions m and n are usually defined as shown in Fig. 3.6. (If the base plate is small, the area of the base plate inside the column profile should be treated as a beam.) Yield-line analysis shows that an equivalent cantilever dimension n' can be defined as $n' = \frac{1}{4}\sqrt{db_f}$, and the required base plate thickness t_p can be calculated from

$$t_p = 2l\sqrt{\frac{f_p}{F_y}}$$

where $l = \max(m, n, n')$, in (mm)

$f_p = P/(BN) \leq F_p$, ksi (MPa)

F_y = yield strength of base plate, ksi (MPa)

P = column axial load, kip (N)

For columns subjected only to direct load, the welds of column to base plate, as shown in Fig. 3.6, are required principally for withstanding erection stresses. For columns subjected to uplift, the welds must be proportioned to resist the forces.

COMPOSITE COLUMNS

The AISC load-and-resistance factor design (LRFD) specification for structural steel buildings contains provisions for design of concrete-encased compression members. It sets the following requirements for qualification as a composite column: The cross-sectional area of the steel core—shapes, pipe, or tubing—should be at least 4 percent of the total composite area. The concrete should be reinforced with longitudinal load-carrying bars, continuous at framed levels, and lateral ties and other longitudinal bars to restrain the concrete; all should have at least 1 ½ in (38.1 mm) of clear concrete cover. The cross-sectional area of transverse and longitudinal reinforcement should be at least 0.007 in^2 (4.5 mm^2) per in (mm) of bar spacing. Spacing of ties should not exceed two-thirds of the smallest dimension of the composite section. Strength of the concrete f'_c should be between 3 and 8 ksi (20.7 and 55.2 MPa) for normal-weight concrete and at least 4 ksi (27.6 MPa) for light-weight concrete. Specified minimum yield stress F_y of steel core and reinforcement should not exceed 60 ksi (414 MPa). Wall thickness of steel pipe or tubing filled with concrete should be at least $b\sqrt{F_y/3E}$ or $D\sqrt{F_y/8E}$, where b is the width of the face of a rectangular section, D is the outside diameter of a circular section, and E is the elastic modulus of the steel.

The AISC LRFD specification gives the design strength of an axially loaded composite column as ϕP_n, where ϕ = 0.85 and P_n is determined from

$$\phi P_n = 0.85 A_s F_{cr}$$

For $\lambda_c \leq 1.5$

$$F_{cr} = 0.658^{\lambda_c^2} F_{my}$$

For $\lambda_c > 1.5$

$$F_{cr} = \frac{0.877}{\lambda_c^2} F_{my}$$

where $\lambda_c = (KL/r_m\pi)\sqrt{F_{my}/E_m}$

KL = effective length of column in (mm)

A_s = gross area of steel core in^2 (mm^2)

$F_{my} = F_y + c_1F_{yr}(A_r/A_s) + c_2f_c'(A_c/A_s)$

$E_m = E + c_3E_c(A_c/A_s)$

r_m = radius of gyration of steel core, in $\leq$ 0.3 of the overall thickness of the composite cross section in the plane of buckling for steel shapes

A_c = cross-sectional area of concrete in^2 (mm^2)

A_r = area of longitudinal reinforcement in^2 (mm^2)

E_c = elastic modulus of concrete ksi (MPa)

F_{yr} = specified minimum yield stress of longitudinal reinforcement, ksi (MPa)

For concrete-filled pipe and tubing, $c_1 = 1.0$, $c_2 = 0.85$, and $c_3 = 0.4$. For concrete-encased shapes, $c_1 = 0.7$, $c_2 = 0.6$, and $c_3 = 0.2$.

When the steel core consists of two or more steel shapes, they should be tied together with lacing, tie plates, or batten plates to prevent buckling of individual shapes before the concrete attains $0.75 f_c'$.

The portion of the required strength of axially loaded encased composite columns resisted by concrete should be developed by direct bearing at connections or shear connectors can be used to transfer into the concrete the load applied directly to the steel column. For direct bearing, the design strength of the concrete is $1.7\phi_c f_c' A_b$, where $\phi_c = 0.65$ and A_b = loaded area, in^2 (mm^2). Certain restrictions apply.

ELASTIC FLEXURAL BUCKLING OF COLUMNS

Elastic buckling is a state of lateral instability that occurs while the material is stressed below the yield point. It is of special importance in structures with slender members. Euler's formula for pin-ended columns (Fig. 3.7) gives valid results for the critical buckling load, kip (N). This formula is, with L/r as the slenderness ratio of the column,

$$P = \frac{\pi^2 EA}{(L/r)^2}$$

where E = modulus of elasticity of the column material, psi (Mpa)

A = column cross-sectional area, in^2 (mm^2)

r = radius of gyration of the column, in (mm)

Figure 3.8 shows some ideal end conditions for slender columns and corresponding critical buckling loads. Elastic critical buckling loads may be obtained for all cases by substituting an effective length KL for the length L of the pinned column, giving

$$P = \frac{\pi^2 EA}{(KL/r)^2}$$

In some cases of columns with open sections, such as a cruciform section, the controlling buckling mode may be one of twisting instead of lateral deformation. If the warping rigidity of the section is negligible, *torsional buckling* in a pin-ended column occurs at an axial load of

$$P = \frac{GJA}{I_p}$$

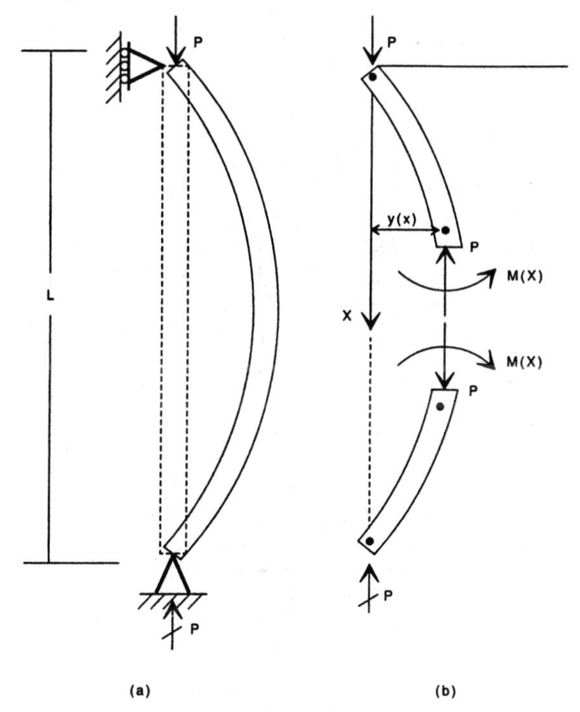

FIGURE 3.7 (*a*) Buckling of a pin-ended column under axial load. (*b*) Internal forces hold the column in equilibrium.

where G = shear modulus of elasticity

J = torsional constant

A = cross-sectional area

I_p = polar moment of inertia = $I_x + I_y$

Type of column	Effective length	Critical buckling load
L	L	$\frac{\pi^2 EI}{L^2}$
L/4 L/2 L/4	$\frac{L}{2}$	$\frac{4\pi^2 EI}{L^2}$
0.3L 0.7L	$\approx 0.7L$	$\approx \frac{2\pi^2 EI}{L^2}$
L	$2L$	$\frac{\pi^2 EI}{4L^2}$

FIGURE 3.8 Buckling formulas for columns.

If the section possesses a significant amount of warping rigidity, the axial buckling load is increased to

$$P = \frac{A}{I_p}\left(GJ + \frac{\pi^2 E C_w}{L^2}\right)$$

where C_w is the warping constant, a function of cross-sectional shape and dimensions.

ALLOWABLE DESIGN LOADS FOR ALUMINUM COLUMNS

Euler's equation is used for long aluminum columns, and depending on the material, either Johnson's parabolic or straight-line equation is used for short columns. These equations for aluminum follow:

Euler's equation:

$$F_e = \frac{c\pi^2 E}{(L/\rho)^2}$$

Johnson's generalized equation:

$$F_c = F_{ce}\left[1 - K\left(\frac{(L/\rho)}{\pi\sqrt{\dfrac{cE}{F_{ce}}}}\right)^n\right]$$

The value of n, which determines whether the short column formula is the straight-line or parabolic type, is selected from Table 3.5. The transition from the long to the short column range is given by

$$\left(\frac{L}{\rho}\right)_{\rm cr} = \pi\sqrt{\frac{kcE}{F_{\rm ce}}}$$

where F_e = allowable column compressive stress

$F_{\rm ce}$ = column yield stress and is given as a function of $F_{\rm cy}$ (compressive yield stress)

L = length of column

ρ = radius of gyration of column

E = modulus of elasticity—noted on nomograms

c = column-end fixity from Fig. 3.9

n, K, k = constants from Table 3.5

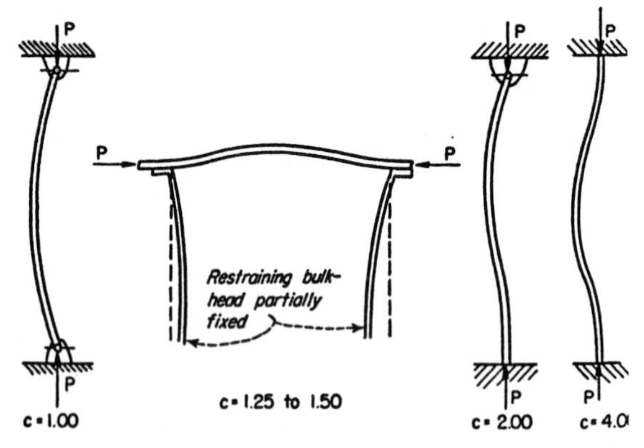

FIGURE 3.9 Values of c, column-end fixity, for determining the critical L/ρ ratio of different loading conditions.

TABLE 3.5 Material Constants for Common Aluminum Alloys

	Average F_{cy}		Values F_{ce}					
Material	psi	MPa	psi	MPa	K	k	n	Type Johnson equation
14S–T4	34,000	234.4	39,800	274.4	0.385	3.00	1.0	Straight line
24S–T3 and T4	40,000	275.8	48,000	330.9	0.385	3.00	1.0	Straight line
61S–T6	35,000	241.3	41,100	283.4	0.385	3.00	1.0	Straight line
14S–T6	57,000	393.0	61,300	422.7	0.250	2.00	2.0	Squared parabolic
75S–T6	69,000	475.8	74,200	511.6	0.250	2.00	2.0	Squared parabolic

Ref: ANC-5.

ULTIMATE STRENGTH DESIGN CONCRETE COLUMNS

At ultimate strength P_u, kip (N), columns should be capable of sustaining loads as given by the American Concrete Institute *required strength* equations in Chap. 5, "Concrete Formulas" at actual eccentricities. P_u, may not exceed ϕP_n, where ϕ is the capacity reduction factor and P_n, kip (N), is the column ultimate strength. If P_o, kip (N), is the column ultimate strength with zero eccentricity of load, then

$$P_o = 0.85 f'_c (A_g - A_{st}) + f_y A_{st}$$

where f_y = yield strength of reinforcing steel, ksi (MPa)

f'_c = 28-day compressive strength of concrete, ksi (MPa)

A_g = gross area of column, in^2 (mm^2)

A_{st} = area of steel reinforcement, in^2 (mm^2)

For members with spiral reinforcement then, for axial loads only,

$$P_u \leq 0.85 \phi P_o$$

For members with tie reinforcement, for axial loads only,

$$P_u \leq 0.80 \phi P_o$$

Eccentricities are measured from the plastic centroid. This is the centroid of the resistance to load computed for the assumptions that the concrete is stressed uniformly to $0.85 f'_c$ and the steel is stressed uniformly to f_y.

The axial-load capacity P_u kip (N), of short, rectangular members subject to axial load and bending may be determined from

$$P_u = \phi(0.85 f'_c ba + A'_s f_y - A_s f_s)$$

$$P_u e' = \phi\left[0.85 f'_c ba\left(d - \frac{a}{2}\right) + A'_s f_y (d - d')\right]$$

where e' = eccentricity, in (mm), of axial load at end of member with respect to centroid of tensile reinforcement, calculated by conventional methods of frame analysis

b = width of compression face, in (mm)

a = depth of equivalent rectangular compressive-stress distribution, in (mm)

A'_s = area of compressive reinforcement, in^2 (mm^2)

A_s = area of tension reinforcement, in^2 (mm^2)

d = distance from extreme compression surface to centroid of tensile reinforcement, in (mm)

d' = distance from extreme compression surface to centroid of compression reinforcement, in (mm)

f_s = tensile stress in steel, ksi (MPa)

The two preceding equations assume that a does not exceed the column depth, that reinforcement is in one or two faces parallel to axis of bending, and that reinforcement in any face is located at about the same distance from the axis of bending. Whether the compression steel actually yields at ultimate strength, as assumed in these and the following equations, can be verified by strain compatibility calculations. That is, when the concrete crushes, the strain in the compression steel, $0.003\,(c - d')/c$, must be larger than the strain when the steel starts to yield, f_y/E_s. In this case, c is the distance, in (mm), from the extreme compression surface to the neutral axis and E_s is the modulus of elasticity of the steel, ksi (MPa).

The load P_b for balanced conditions can be computed from the preceding P_u equation with $f_s = f_y$ and

$$\begin{aligned} a &= a_b \\ &= \beta_1 c_b \\ &= \frac{87{,}000\ \beta_1 d}{87{,}000 + f_y} \end{aligned}$$

The balanced moment, in. · kip (k · Nm), can be obtained from

$$\begin{aligned} M_b &= P_b e_b \\ &= \phi\left[0.85 f_c'\ b a_b\left(d - d'' - \frac{a_b}{2}\right)\right. \\ &\qquad \left. + A_s' f_y (d - d' - d'') + A_s f_y d''\right] \end{aligned}$$

where e_b is the eccentricity, in (mm), of the axial load with respect to the plastic centroid and d'' is the distance, in (mm), from plastic centroid to centroid of tension reinforcement.

When P_u is less than P_b or the eccentricity, e, is greater than e_b, tension governs. In that case, for unequal tension and compression reinforcement, the ultimate strength is

$$\begin{aligned} P_u = 0.85 f_c' b d \phi &\left\{\rho' m' - \rho m + \left(1 - \frac{e'}{d}\right)\right. \\ &\left. + \sqrt{\left(1 - \frac{e'}{d}\right)^2 + 2\left[(\rho m - \rho' m')\frac{e'}{d} + \rho' m'\left(1 - \frac{d'}{d}\right)\right]}\right\} \end{aligned}$$

where
$$m = f_y'/0.85f_c'$$
$$m' = m - 1$$
$$\rho = A_s/bd$$
$$\rho' = A_s'/bd$$

Special Cases of Reinforcement

For symmetrical reinforcement in two faces, the preceding P_u equation becomes

$$P_u = 0.85\, f_c' bd\phi \left\{ -\rho + 1 - \frac{e'}{d} + \sqrt{\left(1 - \frac{e'}{d}\right)^2 + 2\rho\left[m'\left(1 - \frac{d'}{d}\right) + \frac{e'}{d}\right]} \right\}$$

Column Strength When Compression Governs

For no compression reinforcement, the P_u equation becomes

$$P_u = 0.85\, f_c' bd\phi \left[-\rho m + 1 - \frac{e'}{d} + \sqrt{\left(1 - \frac{e'}{d}\right)^2 + 2\frac{e'\rho m}{d}} \right]$$

When P_u is greater than P_b, or e is less than e_b, compression governs. In that case, the ultimate strength is approximately

$$P_u = P_o - (P_o - P_b)\frac{M_u}{M_b}$$

$$P_u = \frac{P_o}{1 + (P_o/P_b - 1)(e/e_b)}$$

where M_u is the moment capacity under combined axial load and bending, in kip (k Nm) and P_o is the axial-load capacity, kip (N), of member when concentrically loaded, as given.

For symmetrical reinforcement in single layers, the ultimate strength when compression governs in a column with depth, h, may be computed from

$$P_u = \phi\left(\frac{A_s' f_y}{e/d - d' + 0.5} + \frac{bhf_c'}{3he/d^2 + 1.18}\right)$$

Circular Columns

Ultimate strength of short, circular members with bars in a circle may be determined from the following equations:

When tension controls,

$$P_u = 0.85 f_c' D^2 \phi \left[\sqrt{\left(\frac{0.85e}{D} - 0.38\right)^2 + \frac{\rho_t m D_s}{2.5D}} - \left(\frac{0.85e}{D} - 0.38\right)\right]$$

where D = overall diameter of section, in (mm)

D_s = diameter of circle through reinforcement, in (mm)

$\rho_t = A_{st}/A_g$

When compression governs,

$$P_u = \phi \left[\frac{A_{st} f_v}{3e/D_s + 1} + \frac{A_g f'_c}{9.6De/(0.8D + 0.67D_s)^2 + 1.18} \right]$$

The eccentricity for the balanced condition is given approximately by

$$e_b = (0.24 - 0.39\,\rho_t m)D$$

Short Columns

Ultimate strength of short, square members with depth, h, and with bars in a circle may be computed from the following equations:

When tension controls,

$$P_u = 0.85bhf'_c\phi \left[\sqrt{\left(\frac{e}{h} - 0.5\right)^2 + 0.67\frac{D_s}{h}\rho_t m} - \left(\frac{e}{h} - 0.5\right) \right]$$

When compression governs,

$$P_u = \phi \left[\frac{A_{st} f_y}{3e/D_s + 1} + \frac{A_g f'_c}{12he/(h + 0.67D_s)^2 + 1.18} \right]$$

Slender Columns

When the slenderness of a column has to be taken into account, the eccentricity should be determined from $e = M_c/P_u$, where M_c is the magnified moment.

CHAPTER 4

PILES AND PILING FORMULAS

ALLOWABLE LOADS ON PILES

A dynamic formula extensively used in the United States to determine the allowable static load on a pile is the *Engineering News* formula. For piles driven by a drop hammer, the allowable load is

$$P_a = \frac{2WH}{p + 1}$$

For piles driven by a steam hammer, the allowable load is

$$P_a = \frac{2WH}{p + 0.1}$$

where P_a = allowable pile load, tons (kg)

W = weight of hammer, tons (kg)

H = height of drop, ft (m)

p = penetration of pile per blow, in (mm)

The preceding two equations include a factor of safety of 6.

For a group of piles penetrating a soil stratum of good bearing characteristics and transferring their loads to the soil by point bearing on the ends of the piles, the total allowable load would be the sum of the individual allowable loads for each pile. For piles transferring their loads to the soil by skin friction on the sides of the piles, the total allowable load would be less than the sum on the individual allowable loads for each pile, because of the interaction of the shearing stresses and strains caused in the soil by each pile.

LATERALLY LOADED VERTICAL PILES

Vertical-pile resistance to lateral loads is a function of both the flexural stiffness of the shaft, the stiffness of the bearing soil in the upper 4 to $6D$ length of shaft, where D = pile diameter and the degree of pile-head fixity.

The lateral-load vs. pile-head deflection relationship is developed from charted nondimensional solutions of Reese and Matlock. The solution assumes the soil modulus K to increase linearly with depth z; that is, $K = n_h z$, where n_h = coefficient of horizontal subgrade reaction. A characteristic pile length T is calculated from

$$T = \sqrt{\frac{EI}{n_h}}$$

where EI = pile stiffness. The lateral deflection y of a pile with head free to move and subject to a lateral load P_t and moment M_t applied at the ground line is given by

$$y = A_y P_t \frac{T^3}{EI} + B_y M_t \frac{T^2}{EI}$$

where A_y and B_y are nondimensional coefficients. Nondimensional coefficients are also available for evaluation of pile slope, moment, shear, and soil reaction along the shaft.

For positive moment,

$$M = A_m P_t T + B_m M_t$$

Positive M_t and P_t values are represented by clockwise moment and loads directed to the right on the pile head at the ground line. The coefficients applicable to evaluation of pile-head deflection and to the maximum positive moment and its approximate position on the shaft, z/T, where z = distance below the ground line, are listed in Table 4.1.

TABLE 4.1 Percentage of Base Load Transmitted to Rock Socket

	E_r/E_p		
L_s/d_s	0.25	1.0	4.0
0.5	54†	48	44
1.0	31	23	18
1.5	17†	12	8†
2.0	13†	8	4

†Estimated by interpretation of finite-element solution; for Poisson's ratio = 0.26.

The negative moment imposed at the pile head by pile-cap or another structural restraint can be evaluated as a function of the head slope (rotation) from

$$-M_t = \frac{A_\theta P_t T}{B_\theta} - \frac{\theta_s EI}{B_\theta T}$$

where θ_s rad represents the counterclockwise (+) rotation of the pile head and A_θ and B_θ are coefficients (see Table 4.1). The influence of the degrees of fixity of the pile head on y and M can be evaluated by substituting the value of $-M_t$ from the preceding equation into the earlier y and M equations. Note that, for the fixed-head case,

$$y_f = \frac{P_t T^3}{EI}\left(A_y - \frac{A_\theta B_y}{B_\theta}\right)$$

TOE CAPACITY LOAD

For piles installed in cohesive soils, the ultimate tip load may be computed from

$$Q_{\text{bu}} = A_b q = A_b N_c c_u \tag{4.1}$$

where A_b = end-bearing area of pile

q = bearing capacity of soil

N_t = bearing-capacity factor

c_u = undrained shear strength of soil within zone 1 pile diameter above and 2 diameters below pile tip

Although theoretical conditions suggest that N_c may vary between about 8 and 12, N_c is usually taken as 9.

For cohesionless soils, the toe resistance stress, q, is conventionally expressed by Eq. (4.1) in terms of a bearing-capacity factor N_q and the effective overburden pressure at the pile tip σ'_{vo}:

$$q = N_q \sigma'_{\text{vo}} \leq q_l \tag{4.2}$$

Some research indicates that, for piles in sands, q, like $\bar{f}_s$, reaches a quasi-constant value, q_l, after penetrations of the bearing stratum in the range of 10 to 20 pile diameters. Approximately:

$$q_l = 0.5 N_q \tan \phi \tag{4.3}$$

where ϕ is the friction angle of the bearing soils below the critical depth. Values of N_q applicable to piles are given in Fig. 4.1. Empirical correlations of soil test data with q and q_l have also been applied to predict successfully end-bearing capacity of piles in sand.

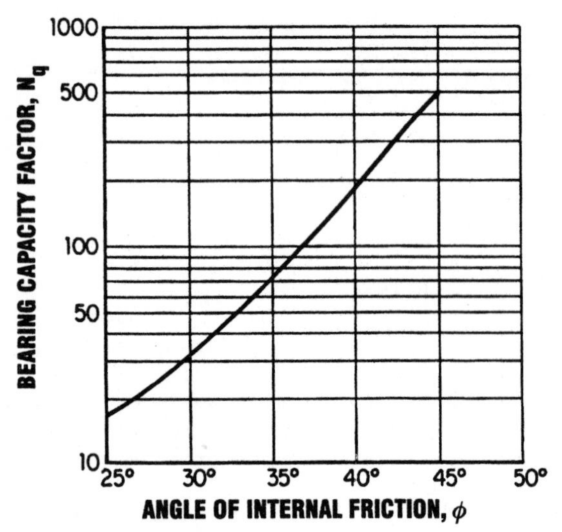

FIGURE 4.1 Bearing-capacity factor for granular soils related to angle of internal friction.

GROUPS OF PILES

A pile group may consist of a cluster of piles or several piles in a row. The group behavior is dictated by the group geometry and the direction and location of the load, as well as by subsurface conditions.

Ultimate-load considerations are usually expressed in terms of a group efficiency factor, which is used to reduce the capacity of each pile in the group. The efficiency factor E_g is defined as the ratio of the ultimate group capacity to the sum of the ultimate capacity of each pile in the group.

E_g is conventionally evaluated as the sum of the ultimate peripheral friction resistance and end-bearing capacities of a block of soil with breadth B, width W, and length L, approximately that of the pile group. For a given pile, spacing S and number of piles n,

$$E_g = \frac{2(BL + WL)\bar{f}_s + BW_g}{nQ_u} \tag{4.4}$$

where $\bar{f}_s$ is the average peripheral friction stress of block and Q_u is the single-pile capacity. The limited number of pile-group tests and model tests available suggest that for cohesive soils $E_g > 1$ if S is more than 2.5 pile diameters D and for cohesionless soils $E_g > 1$ for the smallest practical spacing. A possible exception might be for very short, heavily tapered piles driven in very loose sands.

In practice, the *minimum pile spacing* for conventional piles is in the range of 2.5 to 3.0D. A larger spacing is typically applied for expanded-base piles.

A very approximate method of pile-group analysis calculates the upper limit of group drag load, Q_{gd} from

$$Q_{\text{gd}} = A_F \gamma_F H_F + PHc_u \tag{4.5}$$

where H_f, γ_f, and A_F represent the thickness, unit weight, and area of fill contained within the group. P, H, and c_u are the circumference of the group, the thickness of the consolidating soil layers penetrated by the piles, and their undrained shear strength, respectively. Such forces as Q_{gd} could only be approached for the case of piles driven to rock through heavily surcharged, highly compressible subsoils.

Design of rock sockets is conventionally based on

$$Q_d = \pi d_s L_s f_R + \frac{\pi}{4} d_s^2 q_a \qquad (4.6)$$

where Q_d = allowable design load on rock socket

d_s = socket diameter

L_s = socket length

f_R = allowable concrete-rock bond stress

q_a = allowable bearing pressure on rock

Load-distribution measurements show, however, that much less of the load goes to the base than is indicated by Eq. (4.6). This behavior is demonstrated by the data in Table 4.1, where L_s/d_s is the ratio of the shaft length to shaft diameter and E_r/E_p is the ratio of rock modulus to shaft modulus. The finite-element solution summarized in Table 4.1 probably reflects a realistic trend if the average socket-wall shearing resistance does not exceed the ultimate f_R value; that is, slip along the socket side-wall does not occur.

A simplified design approach, taking into account approximately the compatibility of the socket and base resistance, is applied as follows:

1. Proportion the rock socket for design load Q_d with Eq. (4.6) on the assumption that the end-bearing stress is less than q_a [say $q_a/4$, which is equivalent to assuming that the base load $Q_b = (\pi/4)\ d_s^2 q_a/4$].
2. Calculate $Q_b = RQ_d$, where R is the base-load ratio interpreted from Table 4.1.
3. If RQ_d does not equal the assumed Q_b, repeat the procedure with a new q_a value until an approximate convergence is achieved and $q \leq q_a$.

The final design should be checked against the established settlement tolerance of the drilled shaft.

Following the recommendations of Rosenberg and Journeaux, a more realistic solution by the previous method is obtained if f_{Ru} is substituted for f_R. Ideally, f_{Ru} should be determined from load tests. If this parameter is selected from data that are not site specific, a safety factor of at least 1.5 should be applied to f_{Ru} in recognition of the uncertainties associated with the *UC* strength correlations (Rosenberg, P. and Journeaux, N. L., "Friction and End-Bearing Tests on Bedrock for High-Capacity Socket Design," *Canadian Geotechnical Journal*, 13(3)).

FOUNDATION-STABILITY ANALYSIS

The maximum load that can be sustained by shallow foundation elements at incipient failure (*bearing capacity*) is a function of the cohesion and friction angle of bearing soils as well as the width B and shape of the foundation. The *net bearing capacity* per unit area, q_u, of a long footing is conventionally expressed as

$$q_u = \alpha_f c_u N_c + \sigma'_{\text{vo}} N_q + \beta_f \gamma B N_\gamma \qquad (4.7)$$

where α_f = 1.0 for strip footings and 1.3 for circular and square footings

c_u = undrained shear strength of soil

σ'_{vo} = effective vertical shear stress in soil at level of bottom of footing

β_f = 0.5 for strip footings, 0.4 for square footings, and 0.6 for circular footings

γ = unit weight of soil

B = width of footing for square and rectangular footings and radius of footing for circular footings

N_c, N_q, N_γ = bearing-capacity factors, functions of angle of internal friction ϕ

For undrained (rapid) loading of cohesive soils, $\phi = 0$ and Eq. (4.7) reduces to

$$q_u = N_c' \, c_u \tag{4.8}$$

where $N_c' = \alpha_f N_c$. For drained (slow) loading of cohesive soils, ϕ and c_u are defined in terms of effective friction angle ϕ' and effective stress c_u'.

Modifications of Eq. (4.7) are also available to predict the bearing capacity of layered soil and for eccentric loading.

Rarely, however, does q_u control foundation design when the safety factor is within the range of 2.5 to 3. (Should creep or local yield be induced, excessive settlements may occur. This consideration is particularly important when selecting a safety factor for foundations on soft to firm clays with medium to high plasticity.)

Equation (4.7) is based on an infinitely long strip footing and should be corrected for other shapes. Correction factors by which the bearing-capacity factors should be multiplied are given in Table 4.2, in which L = footing length.

The derivation of Eq. (4.7) presumes the soils to be homogeneous throughout the stressed zone, which is seldom the case. Consequently, adjustments may be required for departures from homogeneity. In sands, if there is a moderate variation in strength, it is safe to use Eq. (4.7), but with bearing-capacity factors representing a weighted average strength.

TABLE 4.2 Shape Corrections for Bearing-Capacity Factors of Shallow Foundations[†]

Shape of foundation	Correction factor		
	N_c	N_q	N_y
Rectangle[‡]	$1 + \left(\frac{B}{L}\right)\left(\frac{N_q}{N_c}\right)$	$1 + \left(\frac{B}{L}\right) \tan \phi$	$1 - 0.4\left(\frac{B}{L}\right)$
Circle and square	$1 + \left(\frac{N_q}{N_c}\right)$	$1 + \tan \phi$	0.60

[†]After De Beer, E. E., as modified by Vesic, A. S. See Fang, H. Y., *Foundation Engineering Handbook*, 2d ed., Van Nostrand Reinhold, New York.
[‡]No correction factor is needed for long-strip foundations.

Eccentric loading can have a significant impact on selection of the bearing value for foundation design. The conventional approach is to proportion the foundation to maintain the resultant force within its middle third. The footing is assumed to be rigid and the bearing pressure is assumed to vary linearly as shown by Fig. (4.2*b*.) If the resultant lies outside the middle third of the footing, it is assumed that there is bearing over only a portion of the footing, as shown in Fig. (4.2*d*.) For the conventional case, the maximum and minimum bearing pressures are

$$q_m = \frac{P}{BL}\left(1 \pm \frac{6e}{B}\right) \tag{4.9}$$

where B = width of rectangular footing

L = length of rectangular footing

e = eccentricity of loading

For the other case (Fig. 4.3*c*), the soil pressure ranges from 0 to a maximum of

$$q_m = \frac{2P}{3L(B/2 - e)} \tag{4.10}$$

For square or rectangular footings subject to overturning about two principal axes and for unsymmetrical footings, the loading eccentricities e_1 and e_2 are determined about the two principal axes. For the case where the full bearing area of the footings is engaged, q_m is given in terms of the distances from the principal axes, c_1 and c_2, the radius of gyration of the footing area about the principal axes r_1 and r_2, and the area of the footing A as

$$q_m = \frac{P}{A}\left(1 + \frac{e_1 c_1}{r_1^2} + \frac{e_2 c_2}{r_2^2}\right) \tag{4.11}$$

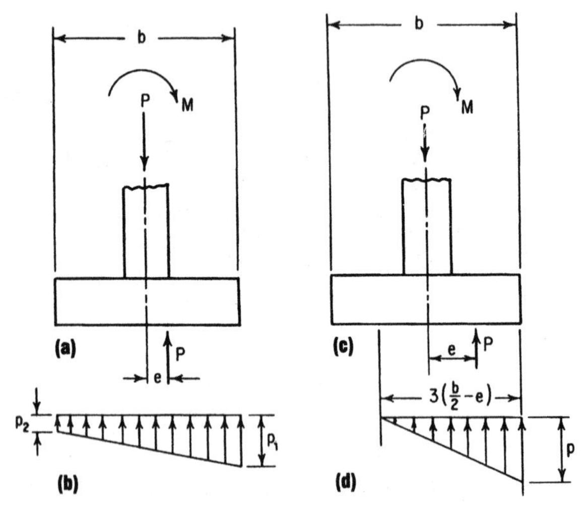

FIGURE 4.2 Footings subjected to overturning.

For the case where only a portion of the footing is bearing, the maximum pressure may be approximated by trial and error.

For all cases of *sustained eccentric loading*, the maximum (edge) pressures should not exceed the shear strength of the soil and also the factor of safety should be at least 1.5 (preferably 2.0) against overturning.

AXIAL-LOAD CAPACITY OF SINGLE PILES

Pile capacity Q_u may be taken as the sum of the shaft and toe resistances, Q_{su} and Q_{bu}, respectively.

The allowable load Q_a may then be determined from either Eq. (4.12) or (4.13):

$$Q_a = \frac{Q_{\mathrm{su}} + Q_{\mathrm{bu}}}{F} \tag{4.12}$$

$$Q_a = \frac{Q_{\mathrm{su}}}{F_1} + \frac{Q_{\mathrm{bu}}}{F_2} \tag{4.13}$$

where F, F_1, and F_2 are safety factors. Typically, F for permanent structures is between 2 and 3, but may be larger, depending on the perceived reliability of the analysis and construction as well as the consequences of failure. Equation (4.13) recognizes that the deformations required to fully mobilize Q_{su} and Q_{bu} are not compatible. For example, Q_{su} may be developed at displacements less than 0.25 in (6.35 mm), whereas Q_{bu} may be realized at a toe displacement equivalent to 5 percent to 10 percent of the pile diameter. Consequently, F_1 may be taken as 1.5 and F_2 as 3.0, if the equivalent single safety factor equals F or larger. (If $Q_{\mathrm{su}}/Q_{\mathrm{bu}} < 1.0$, F is less than the

2.0 usually considered as a major safety factor for permanent structures.)

SHAFT SETTLEMENT

Drilled-shaft settlements can be estimated by empirical correlations or by load-deformation compatibility analyses. Other methods used to estimate settlement of drilled shafts, singly or in groups, are identical to those used for piles. These include elastic, semiempirical elastic, and load-transfer solutions for single shafts drilled in cohesive or cohesionless soils.

Resistance to tensile and lateral loads by straight-shaft drilled shafts should be evaluated as described for pile foundations. For relatively rigid shafts with characteristic length T greater than 3, there is evidence that bells increase the lateral resistance. The added ultimate resistance to uplift of a belled shaft Q_{ut} can be approximately evaluated for cohesive soils models for bearing capacity [Eq. (4.14)] and friction cylinder [Eq. (4.15)] as a function of the shaft diameter D and bell diameter D_b (Meyerhof, G. G. and Adams, J. I., "The Ultimate Uplift Capacity of Foundations," *Canadian Geotechnical Journal*, 5(4):1968.)

For the bearing-capacity solution,

$$Q_{ul} = \frac{\pi}{4}(D_b^2 - D^2)N_c\,\omega c_u + W_p \qquad (4.14)$$

The shear-strength reduction factor ω in Eq. (4.14) considers disturbance effects and ranges from $\frac{1}{2}$ (slurry construction) to $\frac{3}{4}$ (dry construction). The c_u represents the undrained shear strength of the soil just above the bell surface, and N_c is a bearing-capacity factor.

The failure surface of the friction cylinder model is conservatively assumed to be vertical, starting from the base of the bell. Q_{ut} can then be determined for both cohesive and cohesionless soils from

$$Q_{ul} = \pi_b L f_{ut} + W_s + W_p \tag{4.15}$$

where f_{ut} is the average ultimate skin-friction stress in tension developed on the failure plane; that is, $f_{ut} = 0.8\bar{c}_u$ for clays or $K\bar{\sigma}'_{vo}$ tan ϕ for sands. W_s and W_p represent the weight of soil contained within the failure plane and the shaft weight, respectively.

SHAFT RESISTANCE IN COHESIONLESS SOILS

The shaft resistance stress $\bar{f}_s$ is a function of the soil-shaft friction angle δ, degree, and an empirical lateral earth-pressure coefficient K:

$$\bar{f}_s = K\bar{\sigma}'_{vo} \tan \delta \leq f_l \tag{4.16}$$

At displacement-pile penetrations of 10 to 20 pile diameters (loose to dense sand), the average skin friction reaches a limiting value f_l. Primarily depending on the relative density and texture of the soil, f_l has been approximated conservatively by using Eq. (4.16) to calculate $\bar{f}_s$.

For relatively long piles in sand, K is typically taken in the range of 0.7 to 1.0 and δ is taken to be about $\phi - 5$, where ϕ is the angle of internal friction, degree. For piles less than 50 ft (15.2 m) long, K is more likely to be in the range of 1.0 to 2.0, but can be greater than 3.0 for tapered piles.

Empirical procedures have also been used to evaluate $\bar{f}_s$ from *in situ* tests, such as cone penetration, standard penetration, and relative density tests. Equation (4.17), based on standard penetration tests, as proposed by Meyerhof, is generally conservative and has the advantage of simplicity:

$$\bar{f}_s = \frac{\overline{N}}{50} \tag{4.17}$$

where $\overline{N}$ = average standard penetration resistance within the embedded length of pile and $\bar{f}_s$ is given in tons/ft^2. (Meyerhof, G. G., "Bearing Capacity and Settlement of Pile Foundations," *ASCE Journal of Geotechnical Engineering Division*, 102(GT3):1976.)

CHAPTER 5
CONCRETE FORMULAS

REINFORCED CONCRETE

When working with reinforced concrete and when designing reinforced concrete structures, the *American Concrete Institute* (ACI) *Building Code Requirements for Reinforced Concrete*, latest edition, is widely used. Future references to this document are denoted as the ACI *Code*. Likewise, publications of the *Portland Cement Association* (PCA) find extensive use in design and construction of reinforced concrete structures.

Formulas in this chapter cover the general principles of reinforced concrete and its use in various structural applications. Where code requirements have to be met, the reader must refer to the current edition of the ACI *Code* previously mentioned. Likewise, the PCA publications should also be referred to for the latest requirements and recommendations.

WATER/CEMENTITIOUS MATERIALS RATIO

The water/cementitious (w/c) ratio is used in both tensile and compressive strength analyses of Portland concrete cement. This ratio is found from

$$\frac{w}{c} = \frac{w_m}{w_c}$$

where w_m = weight of mixing water in batch, lb (kg); and w_c = weight of cementitious materials in batch, lb (kg).

The ACI *Code* lists the typical relationship between the w/c ratio by weight and the compressive strength of concrete. Ratios for non-air-entrained concrete vary between 0.41 for a 28-day compressive strength of 6000 lb/in^2 (41 MPa) and 0.82 for 2000 lb/in^2 (14 MPa). Air-entrained concrete w/c ratios vary from 0.40 to 0.74 for 5000 lb/in^2 (34 MPa) and 2000 lb/in^2 (14 MPa) compressive strength, respectively. Be certain to refer to the ACI *Code* for the appropriate w/c value when preparing designs or concrete analyses.

Further, the ACI *Code* also lists maximum w/c ratios when strength data are not available. Absolute w/c ratios by weight vary from 0.67 to 0.38 for non-air-entrained concrete and from 0.54 to 0.35 for air-entrained concrete. These values are for a specified 28-day compressive strength f'_c in lb/in^2 or MPa, of 2500 lb/in^2 (17 MPa) to 5000 lb/in^2 (34 MPa). Again, refer to the ACI *Code* before making any design or construction decisions.

Maximum w/c ratios for a variety of construction conditions are also listed in the ACI *Code*. Construction conditions include concrete protected from exposure to freezing and thawing; concrete intended to be watertight; and concrete exposed to deicing salts, brackish water, seawater, etc. Application formulas for w/c ratios are given later in this chapter.

JOB MIX CONCRETE VOLUME

A trial batch of concrete can be tested to determine how much concrete is to be delivered by the job mix. To determine the volume obtained for the job, add the *absolute volume* V_a of the four components—cements, gravel, sand, and water.

Find the V_a for each component from

$$V_a = \frac{W_L}{(\mathrm{SG})W_u}$$

where V_a = absolute volume, ft^3 (m^3)

W_L = weight of material, lb (kg)

SG = specific gravity of the material

w_u = density of water at atmospheric conditions (62.4 lb/ft^3; 1000 kg/m^3)

Then, job yield equals the sum of V_a for cement, gravel, sand, and water.

MODULUS OF ELASTICITY OF CONCRETE

The modulus of elasticity of concrete E_c—adopted in modified form by the ACI *Code*—is given by

$$E_c = 33w_c^{1.5}\sqrt{f_c'} \quad \text{lb/in}^2 \text{ in USCS units}$$
$$= 0.043w_c^{1.5}\sqrt{f_c'} \quad \text{MPa in SI units}$$

With normal-weight, normal-density concrete these two relations can be simplified to

$$E_c = 57{,}000\sqrt{f_c'} \quad \text{lb/in}^2 \text{ in USCS units}$$
$$= 4700\sqrt{f_c'} \quad \text{MPa in SI units}$$

where E_c = modulus of elasticity of concrete, lb/in^2 (MPa); and f_c' = specified 28-day compressive strength of concrete, lb/in^2 (MPa).

TENSILE STRENGTH OF CONCRETE

The tensile strength of concrete is used in combined-stress design. In normal-weight, normal-density concrete the tensile strength can be found from

$$f_r = 7.5\sqrt{f_c'} \quad \text{lb/in}^2 \text{ in USCS units}$$

$$f_r = 0.7\sqrt{f_c'} \quad \text{MPa in SI units}$$

REINFORCING STEEL

American Society for Testing and Materials (ASTM) specifications cover renforcing steel. The most important properties of reinforcing steel are

1. Modulus of elasticity E_s, lb/in^2 (MPa)
2. Tensile strength, lb/in^2 (MPa)
3. Yield point stress f_y, lb/in^2 (MPa)
4. Steel grade designation (yield strength)
5. Size or diameter of the bar or wire

CONTINUOUS BEAMS AND ONE-WAY SLABS

The ACI *Code* gives approximate formulas for finding shear and bending moments in continuous beams and one-way slabs. A summary list of these formulas follows. They are equally applicable to USCS and SI units. Refer to the ACI *Code* for specific applications of these formulas.

For Positive Moment

End spans	
If discontinuous end is unrestrained	$wl_n^2/11$
If discontinuous end is integral with the support	$wl_n^2/14$
Interior spans	$wl_n^2/16$

For Negative Moment

Negative moment at exterior face of first interior support	
Two spans	$wl_n^2/9$
More than two spans	$wl_n^2/10$
Negative moment at other faces of interior supports	$wl_n^2/11$
Negative moment at face of all supports for (*a*) slabs with spans not exceeding 10 ft (3 m) and (*b*) beams and girders where the ratio of sum of column stiffness to beam stiffness exceeds 8 at each end of the span	$wl_n^2/12$
Negative moment at interior faces of exterior supports, for members built integrally with their supports	
Where the support is a spandrel beam or girder	$wl_n^2/24$
Where the support is a column	$wl_n^2/16$

Shear Forces

Shear in end members at first interior support	$1.15\ wl_n/2$
Shear at all other supports	$wl_n/2$

End Reactions

Reactions to a supporting beam, column, or wall are obtained as the sum of shear forces acting on both sides of the support.

DESIGN METHODS FOR BEAMS, COLUMNS, AND OTHER MEMBERS

A number of different design methods have been used for reinforced concrete construction. The three most common are *working-stress design, ultimate-strength design*, and *strength design method*. Each method has its backers and supporters. For actual designs the latest edition of the ACI *Code* should be consulted.

Beams

Concrete beams may be considered to be of three principal types: (1) rectangular beams with tensile reinforcing only, (2) T beams with tensile reinforcing only, and (3) beams with tensile and compressive reinforcing.

Rectangular Beams with Tensile Reinforcing Only. This type of beam includes slabs, for which the beam width b equals 12 in (305 mm) when the moment and shear are expressed per foot (m) of width. The stresses in the concrete and steel are, using working-stress design formulas,

$$f_c = \frac{2M}{kjbd^2} \qquad f_s = \frac{M}{A_s jd} = \frac{M}{pjbd^2}$$

where b = width of beam [equals 12 in (304.8 mm) for slab], in (mm)

d = effective depth of beam, measured from compressive face of beam to centroid of tensile reinforcing (Fig. 5.1), in (mm)

M = bending moment, lb · in (k · Nm)

f_c = compressive stress in extreme fiber of concrete, lb/in^2 (MPa)

f_s = stress in reinforcement, lb/in^2 (MPa)

A_s = cross-sectional area of tensile reinforcing, in^2 (mm^2)

j = ratio of distance between centroid of compression and centroid of tension to depth d

k = ratio of depth of compression area to depth d

p = ratio of cross-sectional area of tensile reinforcing to area of the beam ($= A_s/bd$)

For approximate design purposes, j may be assumed to be $^7/_8$ and k, $^1/_3$. For average structures, the guides in Table 5.1 to the depth d of a reinforced concrete beam may be used.

For a balanced design, one in which both the concrete and the steel are stressed to the maximum allowable stress, the following formulas may be used:

$$bd^2 = \frac{M}{K} \qquad K = \frac{1}{2} f_c kj = pf_s j$$

Values of K, k, j, and p for commonly used stresses are given in Table 5.2.

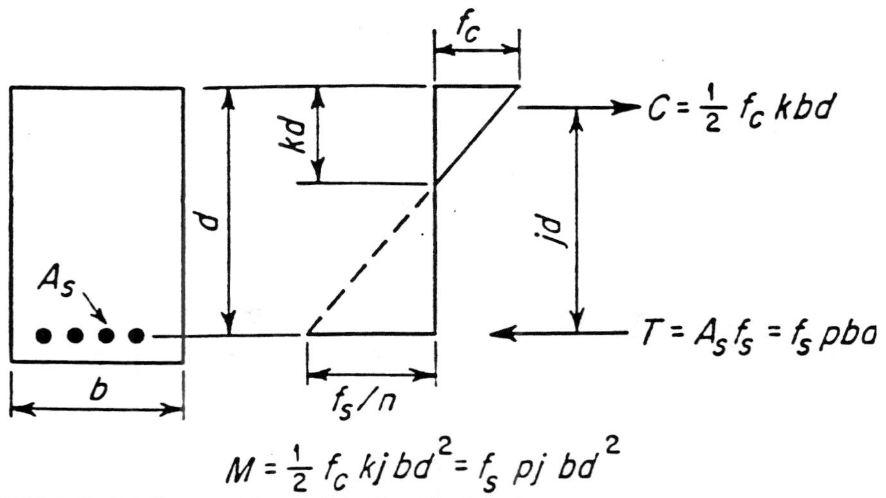

FIGURE 5.1 Rectangular concrete beam with tensile reinforcing only.

TABLE 5.1 Guides to Depth d of Reinforced Concrete Beam†

Member	d
Roof and floor slabs	$l/25$
Light beams	$l/15$
Heavy beams and girders	$l/12$–$l/10$

†l is the span of the beam or slab in inches (millimeters). The width of a beam should be at least $l/32$.

T Beams with Tensile Reinforcing Only. When a concrete slab is constructed monolithically with the supporting concrete beams, a portion of the slab acts as the upper flange of the beam. The effective flange width should not exceed (1) one-fourth the span of the beam, (2) the width of the web portion of the beam plus 16 times the thickness of the slab, or (3) the center-to-center distance between beams. T beams where the upper flange is not a portion of a slab should have a flange thickness not less than one-half the width of the web and a flange width not more than four times the width of the web. For preliminary designs, the preceding formulas given for rectangular beams with tensile reinforcing only can be used, because the neutral axis is usually in, or near, the flange. The area of tensile reinforcing is usually critical.

TABLE 5.2 Coefficients K, k, j, p for Rectangular Sections†

f'_s	n	f_s	K	k	j	p
2000	15	900	175	0.458	0.847	0.0129
2500	12	1125	218	0.458	0.847	0.0161
3000	10	1350	262	0.458	0.847	0.0193
3750	8	1700	331	0.460	0.847	0.0244

†f_s = 16,000 lb/in² (110 MPa).

Beams with Tensile and Compressive Reinforcing. Beams with compressive reinforcing are generally used when the size of the beam is limited. The allowable beam dimensions are used in the formulas given earlier to determine the moment that could be carried by a beam without compressive reinforcement. The reinforcing requirements may then be approximately determined from

$$A_s = \frac{8M}{7f_s d} \qquad A_{sc} = \frac{M - M'}{nf_c d}$$

where A_s = total cross-sectional area of tensile reinforcing, in^2 (mm^2)

A_{sc} = cross-sectional area of compressive reinforcing, in^2 (mm^2)

M = total bending moment, lb·in (K·Nm)

M' = bending moment that would be carried by beam of balanced design and same dimensions with tensile reinforcing only, lb·in (K·Nm)

n = ratio of modulus of elasticity of steel to that of concrete

Checking Stresses in Beams. Beams designed using the preceding approximate formulas should be checked to ensure that the actual stresses do not exceed the allowable, and that the reinforcing is not excessive. This can be accomplished by determining the moment of inertia of the beam. In this determination, the concrete below the neutral axis should not be considered as stressed, whereas the reinforcing steel should be transformed into an equivalent concrete section. For tensile reinforcing, this transformation is made by multiplying the area A_s by n, the ratio of the modulus of elasticity of steel to that of concrete. For compressive reinforcing, the area A_{sc} is multiplied by 2(n – 1). This factor includes allowances for the concrete in compression replaced by the compressive reinforcing and for the plastic flow of concrete. The neutral axis is then located by solving

$$\tfrac{1}{2}bc_c^2 + 2(n - 1)A_{sc}c_{sc} = nA_sc_s$$

for the unknowns c_c, c_{sc}, and c_s (Fig. 5.2). The moment of inertia of the transformed beam section is

$$I = \tfrac{1}{3}bc_c^3 + 2(n - 1)A_{sc}c_{sc}^2 + nA_sc_s^2$$

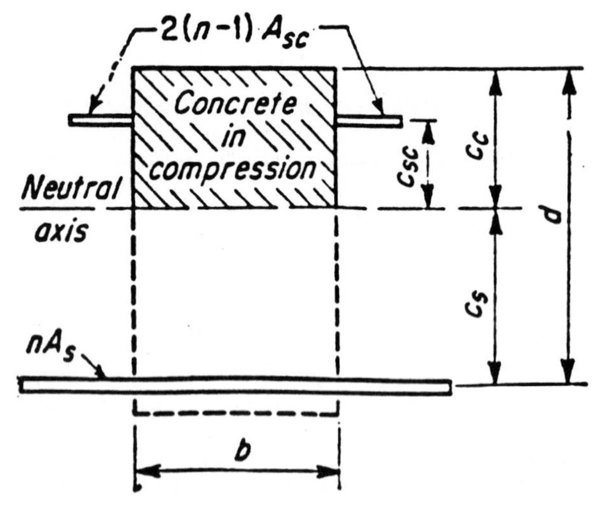

FIGURE 5.2 Transformed section of concrete beam.

and the stresses are

$$f_c = \frac{Mc_c}{I} \qquad f_{sc} = \frac{2nMc_{sc}}{I} \qquad f_s = \frac{nMc_s}{I}$$

where f_c, f_{sc}, f_s = actual unit stresses in extreme fiber of concrete, in compressive reinforcing steel, and in tensile reinforcing steel, respectively, lb/in^2 (MPa)

c_c, c_{sc}, c_s = distances from neutral axis to face of concrete, to compressive reinforcing steel, and to tensile reinforcing steel, respectively, in (mm)

I = moment of inertia of transformed beam section, in^4 (mm^4)

b = beam width, in (mm)

and A_s, A_{sc}, M, and n are as defined earlier in this chapter.

Shear and Diagonal Tension in Beams. The shearing unit stress, as a measure of diagonal tension, in a reinforced concrete beam is

$$v = \frac{V}{bd}$$

where v = shearing unit stress, lb/in^2 (MPa)

V = total shear, lb (N)

b = width of beam (for T beam use width of stem), in (mm)

d = effective depth of beam

If the value of the shearing stress as computed earlier exceeds the allowable shearing unit stress as specified by the ACI *Code*, web reinforcement should be provided. Such reinforcement usually consists of stirrups. The cross-sectional area required for a stirrup placed perpendicular to the longitudinal reinforcement is

$$A_v = \frac{(V - V')s}{f_i d}$$

where A_v = cross-sectional area of web reinforcement in distance s (measured parallel to longitudinal reinforcement), in^2 (mm^2)

f_v = allowable unit stress in web reinforcement, lb/in^2 (MPa)

V = total shear, lb (N)

V' = shear that concrete alone could carry ($= v_c bd$), lb (N)

s = spacing of stirrups in direction parallel to that of longitudinal reinforcing, in (mm)

d = effective depth, in (mm)

Stirrups should be so spaced that every 45° line extending from the middepth of the beam to the longitudinal tension bars is crossed by at least one stirrup. If the total shearing unit stress is in excess of $3\sqrt{f'_c}$ lb/in^2 (MPa), every such line should be crossed by at least two stirrups. The shear stress at any section should not exceed $5\sqrt{f'_c}$ lb/in^2 (MPa).

Bond and Anchorage for Reinforcing Bars. In beams in which the tensile reinforcing is parallel to the compression face, the bond stress on the bars is

TABLE 5.3 Allowable Bond Stresses[†]

	Horizontal bars with more than 12 in (30.5 mm) of concrete cast below the bar[‡]	Other bars[‡]
Tension bars with sizes and deformations conforming to ASTM A305	$\frac{3.4\sqrt{f_c'}}{D}$ or 350, whichever is less	$\frac{4.8\sqrt{f_c'}}{D}$ or 500, whichever is less
Tension bars with sizes and deformations conforming to ASTM A408	$2.1\sqrt{f_c'}$	$3\sqrt{f_c'}$
Deformed compression bars	$6.5\sqrt{f_c'}$ or 400, whichever is less	$6.5\sqrt{f_c'}$ or 400, whichever is less
Plain bars	$1.7\sqrt{f_c'}$ or 160, whichever is less	$2.4\sqrt{f_c'}$ or 160, whichever is less

[†] lb/in^2 (× 0.006895 = MPa).
[‡] f_c' = compressive strength of concrete, lb/in^2 (MPa); D = nominal diameter of bar, in (mm).

$$u = \frac{V}{jd\Sigma_0}$$

where u = bond stress on surface of bar, lb/in^2 (MPa)

V = total shear, lb (N)

d = effective depth of beam, in (mm)

Σ_0 = sum of perimeters of tensile reinforcing bars, in (mm)

For preliminary design, the ratio j may be assumed to be 7/8. Bond stresses may not exceed the values shown in Table 5.3.

Columns

The principal columns in a structure should have a minimum diameter of 10 in (255 mm) or, for rectangular columns, a minimum thickness of 8 in (203 mm) and a minimum gross cross-sectional area of 96 in^2 (61,935 mm^2).

Short columns with closely spaced spiral reinforcing enclosing a circular concrete core reinforced with vertical bars have a maximum allowable load of

$$P = A_g(0.25f_c' + f_s p_g)$$

where P = total allowable axial load, lb (N)

A_g = gross cross-sectional area of column, in^2 (mm^2)

f_c' = compressive strength of concrete, lb/in^2 (MPa)

f_s = allowable stress in vertical concrete reinforcing, lb/in^2 (MPa), equal to 40 percent of the minimum yield strength, but not to exceed 30,000 lb/in^2 (207 MPa)

p_g = ratio of cross-sectional area of vertical reinforcing steel to gross area of column A_g

The ratio p_g should not be less than 0.01 or more than 0.08. The minimum number of bars to be used is six, and the minimum size is No. 5. The spiral reinforcing to be used in a spirally reinforced column is

$$p_s = 0.45\left(\frac{A_g}{A_c} - 1\right)\frac{f'_c}{f_y}$$

where p_s = ratio of spiral volume to concrete-core volume (out-to-out spiral)

A_c = cross-sectional area of column core (out-to-out spiral), in^2 (mm^2)

f_y = yield strength of spiral reinforcement, lb/in^2 (MPa), but not to exceed 60,000 lb/in^2 (413 MPa)

The center-to-center spacing of the spirals should not exceed one-sixth of the core diameter. The clear spacing between spirals should not exceed one-sixth the core diameter, or 3 in (76 mm), and should not be less than 1.375 in (35 mm), or 1.5 times the maximum size of coarse aggregate used.

Short Columns with Ties. The maximum allowable load on short columns reinforced with longitudinal bars and separate lateral ties is 85 percent of that given earlier for spirally reinforced columns. The ratio p_g for a tied column should not be less than 0.01 or more than 0.08. Longitudinal reinforcing should consist of at least four bars; minimum size is No. 5.

Long Columns. Allowable column loads where compression governs design must be adjusted for column length as follows:

1. If the ends of the column are fixed so that a point of contraflexure occurs between the ends, the applied axial load and moments should be divided by R from (R cannot exceed 1.0)

$$R = 1.32 - \frac{0.006h}{r}$$

2. If the relative lateral displacement of the ends of the columns is prevented and the member is bent in a single curvature, applied axial loads and moments should be divided by R from (R cannot exceed 1.0)

$$R = 1.07 - \frac{0.008h}{r}$$

where h = unsupported length of column, in (mm)

r = radius of gyration of gross concrete area, in (mm)

= 0.30 times depth for rectangular column

= 0.25 times diameter for circular column

R = long-column load reduction factor

Applied axial load and moment when tension governs design should be similarly adjusted, except that R varies linearly with the axial load from the values given at the balanced condition.

Combined Bending and Compression. The strength of a symmetrical column is controlled by compression if the equivalent axial load N has an eccentricity e in each principal direction no greater than given by the two following equations and by tension if e exceeds these values in either principal direction.

For spiral columns,

$$e_b = 0.43\, p_g m D_s + 0.14t$$

For tied columns,

$$e_b = (0.67 p_g m + 0.17)d$$

where e = eccentricity, in (mm)

e_b = maximum permissible eccentricity, in (mm)

N = eccentric load normal to cross section of column

p_g = ratio of area of vertical reinforcement to gross concrete area

$m = f_y / 0.85 f_c'$

D_s = diameter of circle through centers of longitudinal reinforcement, in (mm)

t = diameter of column or overall depth of column, in (mm)

d = distance from extreme compression fiber to centroid of tension reinforcement, in (mm)

f_y = yield point of reinforcement, lb/in^2 (MPa)

Design of columns controlled by compression is based on the following equation, except that the allowable load N may not exceed the allowable load P, given earlier, permitted when the column supports axial load only:

$$\frac{f_a}{F_a} + \frac{f_{bx}}{F_b} + \frac{f_{by}}{F_b} \leq 1.0$$

where f_a = axial load divided by gross concrete area, lb/in² (MPa)

f_{bx}, f_{by} = bending moment about x and y axes, divided by section modulus of corresponding transformed uncracked section, lb/in² (MPa)

F_b = allowable bending stress permitted for bending alone, lb/in² (MPa)

$F_a = 0.34(1 + p_g m) f'_c$

The allowable bending load on columns controlled by tension varies linearly with the axial load from M_0 when the section is in pure bending to M_b when the axial load is N_b.

For spiral columns,

$$M_0 = 0.12 A_{st} f_y D_s$$

For tied columns,

$$M_0 = 0.40 A_s f_y (d - d')$$

where A_{st} = total area of longitudinal reinforcement, in² (mm²)

f_y = yield strength of reinforcement, lb/in² (MPa)

D_s = diameter of circle through centers of longitudinal reinforcement, in (mm)

A_s = area of tension reinforcement, in^2 (mm^2)

d = distance from extreme compression fiber to centroid of tension reinforcement, in (mm)

N_b and M_b are the axial load and moment at the balanced condition (i.e., when the eccentricity e equals e_b as determined). At this condition, N_b and M_b should be determined from

$$M_b = N_b e_b$$

When bending is about two axes,

$$\frac{M_x}{M_{0x}} + \frac{M_y}{M_{0y}} \leq 1$$

where M_z and M_y are bending moments about the x and y axes, and M_{0x} and M_{0y} are the values of M_0 for bending about these axes.

PROPERTIES IN THE HARDENED STATE

Strength is a property of concrete that nearly always is of concern. Usually, it is determined by the ultimate strength of a specimen in compression, but sometimes flexural or tensile capacity is the criterion. Because concrete usually gains strength over a long period of time, the compressive strength at 28 days is commonly used as a measure of this property.

The 28-day compressive strength of concrete can be estimated from the 7-day strength by a formula proposed by W. A. Slater:

$$S_{28} = S_7 + 30\sqrt{S_7}$$

where S_{28} = 28-day compressive strength, lb/in^2 (MPa), and S_7 = 7-day strength, lb/in^2 (MPa).

Concrete may increase significantly in strength after 28 days, particularly when cement is mixed with fly ash. Therefore, specification of strengths at 56 or 90 days is appropriate in design.

Concrete strength is influenced chiefly by the water/cement ratio; the higher this ratio is, the lower the strength. The relationship is approximately linear when expressed in terms of the variable C/W, the ratio of cement to water by weight. For a workable mix, without the use of water reducing admixtures,

$$S_{28} = 2700 \frac{C}{W} - 760$$

Tensile strength of concrete is much lower than compressive strength and, regardless of the types of test, usually has poor correlation with f_c'. As determined in flexural tests, the tensile strength (modulus of rupture—not the true strength) is about $7\sqrt{f_c'}$ for the higher strength concretes and $10\sqrt{f_c'}$ for the lower strength concretes.

Modulus of elasticity E_c, generally used in design for concrete, is a secant modulus. In ACI 318, "Building Code Requirements for Reinforced Concrete," it is determined by

$$E_c = w^{1.5} 33\sqrt{f_c'}$$

where w = weight of concrete, lb/ft^3 (kg/m^3); and f_c' = specified compressive strength at 28 days, lb/in^2 (MPa). For normal-weight concrete, with w = 145 lb/ft^3 (kg/m^3),

$$E_c = 57{,}000\sqrt{f_c'}$$

The modulus increases with age, as does the strength.

TENSION DEVELOPMENT LENGTHS

For bars and deformed wire in tension, basic development length is defined by the equations that follow. For No. 11 and smaller bars,

$$l_d = \frac{0.04 A_b f_y}{\sqrt{f'_c}}$$

where A_b = area of bar, in^2 (mm^2)

f_y = yield strength of bar steel, lb/in^2 (MPa)

f'_c = 28-day compressive strength of concrete, lb/in^2 (MPa)

However, l_d should not be less than 12 in (304.8 mm), except in computation of lap splices or web anchorage.

For No. 14 bars,

$$l_d = 0.085 \frac{f_y}{\sqrt{f'_c}}$$

For No. 18 bars,

$$l_d = 0.125 \frac{f_y}{\sqrt{f'_c}}$$

and for deformed wire,

$$l_d = 0.03 d_b \frac{f_y - 20{,}000}{\sqrt{f'_c}} \geq 0.02 \frac{A_w}{S_w} \frac{f_y}{\sqrt{f'_c}}$$

where A_w is the area, in^2 (mm^2); and s_w is the spacing, in (mm), of the wire to be developed. Except in computation of lap splices or development of web reinforcement, l_d should not be less than 12 in (304.8 mm).

COMPRESSION DEVELOPMENT LENGTHS

For bars in compression, the basic development length l_d is defined as

$$l_d = \frac{0.02 f_y d_b}{\sqrt{f_c'}} \geq 0.0003 d_b f_y$$

but l_d not be less than 8 in (20.3 cm) or $0.0003 f_y d_b$.

CRACK CONTROL OF FLEXURAL MEMBERS

Because of the risk of large cracks opening up when reinforcement is subjected to high stresses, the ACI *Code* recommends that designs be based on a steel yield strength f_y no larger than 80 ksi (551.6 MPa). When design is based on a yield strength f_y greater than 40 ksi (275.8 MPa), the cross sections of maximum positive and negative moment should be proportioned for crack control so that specific limits are satisfied by

$$z = f_s \sqrt[3]{d_c A}$$

where f_s = calculated stress, ksi (MPa), in reinforcement at service loads

d_c = thickness of concrete cover, in (mm), measured from extreme tension surface to center of bar closest to that surface

A = effective tension area of concrete, in^2 (mm^2) per bar. This area should be taken as that surrounding main tension reinforcement, having the same centroid as that reinforcement, multiplied by the ratio of the area of the largest bar used to the total area of tension reinforcement

These limits are $z \leq 175$ kip/in (30.6 kN/mm) for interior exposures and $z \leq 145$ kip/in (25.3 kN/mm) for exterior exposures. These correspond to limiting crack widths of 0.016 to 0.013 in (0.406 to 0.33 mm), respectively, at the extreme tension edge under service loads. In the equation for z, f_s should be computed by dividing the bending moment by the product of the steel area and the internal moment arm, but f_s may be taken as 60 percent of the steel yield strength without computation.

REQUIRED STRENGTH

For combinations of loads, the ACI *Code* requires that a structure and its members should have the following ultimate strengths (capacities to resist design loads and their related internal moments and forces):

With wind and earthquake loads not applied,

$$U = 1.4D + 1.7L$$

where D = effect of basic load consisting of dead load plus volume change (shrinkage, temperature) and L = effect of live load plus impact.

When wind loads are applied, the largest of the preceeding equation and the two following equations determine the required strength:

$$U = 0.75(1.4D + 1.7L + 1.7W)$$

$$U = 0.9D + 1.3W$$

where W = effect of wind load.

If the structure can be subjected to earthquake forces E, substitute $1.1E$ for W in the preceding equation.

Where the effects of differential settlement, creep, shrinkage, or temperature change may be critical to the structure, they should be included with the dead load D, and the strength should be at least equal to

$$U = 0.75(1.4D + 1.7L) \geq 1.4(D + T)$$

where T = cumulative effects of temperature, creep, shrinkage, and differential settlement.

DEFLECTION COMPUTATIONS AND CRITERIA FOR CONCRETE BEAMS

The assumptions of working-stress theory may also be used for computing deflections under service loads; that is, elastic-theory deflection formulas may be used for reinforced-concrete beams. In these formulas, the *effective moment of inertia* I_c *is given by*

$$I_e = \left(\frac{M_{cr}}{M_a}\right)^3 I_g + \left[1 - \left(\frac{M_{cr}}{M_a}\right)^3\right] I_{cr} \leq I_g$$

where I_g = moment of inertia of the gross concrete section

M_{cr} = cracking moment

M_a = moment for which deflection is being computed

I_{cr} = cracked concrete (transformed) section

If y_t is taken as the distance from the centroidal axis of the gross section, neglecting the reinforcement, to the extreme surface in tension, the cracking moment may be computed from

$$M_{cr} = \frac{f_r I_g}{y_t}$$

with the modulus of rupture of the concrete $f_r = 7.5\sqrt{f_c'}$.

The deflections thus calculated are those assumed to occur immediately on application of load. Additional long-time deflections can be estimated by multiplying the immediate deflection by 2 when there is no compression reinforcement or by $2 - 1.2A_s'/A_s \geq 0.6$, where A_s' is the area of compression reinforcement and A_s is the area of tension reinforcement.

ULTIMATE-STRENGTH DESIGN OF RECTANGULAR BEAMS WITH TENSION REINFORCEMENT ONLY

Generally, the area A_s of tension reinforcement in a reinforced-concrete beam is represented by the ratio $\rho = A_s/bd$, where b is the beam width and d is the distance from extreme compression surface to the centroid of tension reinforcement. At ultimate strength, the steel at a critical section of the beam is at its yield strength f_y if the concrete does not fail in compression first. Total tension in the steel then will be $A_s f_y = \rho f_y bd$. It is opposed, by an equal compressive force:

$$0.85 f_c' ba = 0.85 f_c' b\beta_1 c$$

where f_c' = 28-day strength of the concrete, ksi (MPa)

a = depth of the equivalent rectangular stress distribution

c = distance from the extreme compression surface to the neutral axis

β_1 = a constant

Equating the compression and tension at the critical section yields

$$c = \frac{pf_y}{0.85\beta_1 f_c'} d$$

The criterion for compression failure is that the maximum strain in the concrete equals 0.003 in/in (0.076 mm/mm). In that case,

$$c = \frac{0.003}{f_s/E_s + 0.003} d$$

where f_s = steel stress, ksi (MPa)

E_s = modulus of elasticity of steel

= 29,000 ksi (199.9 GPa)

Balanced Reinforcing

Under balanced conditions, the concrete reaches its maximum strain of 0.003 when the steel reaches its yield strength f_y. This determines the steel ratio for balanced conditions:

$$\rho_b = \frac{0.85\beta_1 f'_c}{f_y} \frac{87{,}000}{87{,}000 + f_y}$$

Moment Capacity

For such underreinforced beams, the bending-moment capacity of ultimate strength is

$$M_u = 0.90[bd^2 f'_c \omega(1 - 0.59\omega)]$$

$$= 0.90\left[A_s f_y \left(d - \frac{a}{2}\right)\right]$$

where $\omega = \rho f_y / f'_c$ and $a = A_s f_y / 0.85 f'_c$.

Shear Reinforcement

The ultimate shear capacity V_n of a section of a beam equals the sum of the nominal shear strength of the concrete V_c and the nominal shear strength provided by the reinforcement V_s; that is, $V_n = V_c + V_s$. The factored shear force V_u on a section should not exceed

$$\phi V_n = \phi(V_c + V_s)$$

where ϕ = capacity reduction factor (0.85 for shear and torsion). Except for brackets and other short cantilevers, the section for maximum shear may be taken at a distance equal to d from the face of the support.

The shear Vc carried by the concrete alone should not exceed $2\sqrt{f_c'}\, b_w d$, where bw is the width of the beam web and d, the depth of the centroid of reinforcement. (As an alternative, the maximum for Vc may be taken as

$$V_c = \left(1.9\sqrt{f_c'} + 2500\rho_w \frac{V_u d}{M_u}\right) b_w d$$

$$\leq 3.5\sqrt{f_c'} b_w d$$

where $\rho_w = A_s/b_w d$ and V_u and M_u are the shear and bending moment, respectively, at the section considered, but M_u should not be less than $V_u d$.)

When V_u is larger than ϕV_c, the excess shear has to be resisted by web reinforcement.

The area of steel required in vertical stirrups, in^2 (mm^2), per stirrup, with a spacing s, in (mm), is

$$A_v = \frac{V_s S}{f_y d}$$

where f_y = yield strength of the shear reinforcement. A_v is the area of the stirrups cut by a horizontal plane. V_s should not exceed $8\sqrt{f_c'} b_w d$ in sections with web reinforcement, and f_y should not exceed 60 ksi (413.7 MPa). Where shear reinforcement is required and is placed perpendicular to the axis of the member, it should not be spaced farther apart than 0.5d, or more than 24 in (609.6 mm) c to c. When V_s exceeds $4\sqrt{f_c'} b_w d$, however, the maximum spacing should be limited to 0.25d.

Alternatively, for practical design, to indicate the stirrup spacing s for the design shear V_u, stirrup area A_v, and geometry of the member b_w and d,

$$s = \frac{A_v \phi f_y d}{V_u - 2\phi\sqrt{f_c'} b_w d}$$

The area required when a single bar or a single group of parallel bars are all bent up at the same distance from the support at angle α with the longitudinal axis of the member is

$$A_v = \frac{V_s}{f_y \sin \alpha}$$

in which V_s should not exceed $3\sqrt{f_c'}\,b_w d$. A_v is the area cut by a plane normal to the axis of the bars. The area required when a series of such bars are bent up at different distances from the support or when inclined stirrups are used is

$$A_v = \frac{V_s s}{(\sin \alpha + \cos \alpha) f_y d}$$

A minimum area of shear reinforcement is required in all members, except slabs, footings, and joists or where V_u is less than $0.5V_c$.

Development of Tensile Reinforcement

At least one-third of the positive-moment reinforcement in simple beams and one-fourth of the positive-moment reinforcement in continuous beams should extend along the same face of the member into the support, in both cases, at least 6 in (152.4 mm) into the support. At simple supports and at points of inflection, the diameter of the reinforcement should be limited to a diameter such that the development length l_d satisfies

$$l_d = \frac{M_n}{V_u} + l_a$$

where M_n = computed flexural strength with all reinforcing steel at section stressed to f_y

V_u = applied shear at section

l_a = additional embedment length beyond inflection point or center of support

At an inflection point, l_a is limited to a maximum of d, the depth of the centroid of the reinforcement, or 12 times the reinforcement diameter.

Hooks on Bars

The basic development length for a hooked bar with $f_y = 60$ ksi (413.7 MPa) is defined as

$$l_{hb} = \frac{1200d_b}{\sqrt{f_c'}}$$

where d_b is the bar diameter, in (mm), and f_c' is the 28-day compressive strength of the concrete, lb/in^2 (MPa).

WORKING-STRESS DESIGN OF RECTANGULAR BEAMS WITH TENSION REINFORCEMENT ONLY

From the assumption that stress varies across a beam section with the distance from the neutral axis, it follows that

$$\frac{n f_c}{f_s} = \frac{k}{1 - k}$$

where n = modular ratio E_s/E_c

E_s = modulus of elasticity of steel reinforcement, ksi (MPa)

E_c = modulus of elasticity of concrete, ksi (MPa)

f_c = compressive stress in extreme surface of concrete, ksi (MPa)

f_s = stress in steel, ksi (MPa)

kd = distance from extreme compression surface to neutral axis, in (mm)

d = distance from extreme compression to centroid of reinforcement, in (mm)

When the steel ratio $\rho = A_s / bd$, where A_s = area of tension reinforcement, in^2 (mm^2), and b = beam width, in (mm), is known, k can be computed from

$$k = \sqrt{2n\rho + (n\rho)^2} - n\rho$$

Wherever positive-moment steel is required, ρ should be at least $200/f_y$, where f_y is the steel yield stress. The distance jd between the centroid of compression and the centroid of tension, in (mm), can be obtained from

$$j = 1 - \frac{k}{3}$$

Allowable Bending Moment

The moment resistance of the concrete, in · kip (k · Nm) is

$$M_c = {}^1\!/_2 f_c kjbd^2 = K_c bd^2$$

where $K_c = {}^1\!/_2\, f_c kj$. The moment resistance of the steel is

$$M_s = f_s A_s jd = f_s \rho jbd^2 = K_s bd^2$$

where $K_s = f_s \rho j$.

Allowable Shear

The nominal unit shear stress acting on a section with shear V is

$$v = \frac{V}{bd}$$

Allowable shear stresses are 55 percent of those for ultimate-strength design. Otherwise, designs for shear by the working-stress and ultimate-strength methods are the same. Except for brackets and other short cantilevers, the section for maximum shear may be taken at a distance d from the face of the support. In working-stress design, the shear stress v_c carried by the concrete alone should not exceed $1.1\sqrt{f_c'}$. (As an alternative, the maximum for v_c may be taken as $\sqrt{f_c'} + 1300\rho Vd/M$, with a maximum of $1.9\sqrt{f_c'}$; f_c' is the 28-day compressive strength of the concrete, lb/in^2 (MPa), and M is the bending moment at the section but should not be less than Vd.)

At cross sections where the torsional stress v_t exceeds $0.825\sqrt{f_c'}$, v_c should not exceed

$$v_c = \frac{1.1\sqrt{f_c'}}{\sqrt{1 + (v_t/1.2v)^2}}$$

The excess shear $v - v_c$ should not exceed $4.4\sqrt{f_c'}$ in sections with web reinforcement. Stirrups and bent bars should be capable of resisting the excess shear $V' = V - v_c bd$.

The area required in the legs of a vertical stirrup, in^2 (mm^2), is

$$A_v = \frac{V_s'}{f_v d}$$

where s = spacing of stirrups, in (mm); and f_v = allowable stress in stirrup steel, (lb/in^2) (MPa).

For a single bent bar or a single group of parallel bars all bent at an angle α with the longitudinal axis at the same distance from the support, the required area is

$$A_v = \frac{V'}{f_v \sin \alpha}$$

For inclined stirrups and groups of bars bent up at different distances from the support, the required area is

$$A_v = \frac{V_s'}{f_v d(\sin \alpha + \cos \alpha)}$$

Stirrups in excess of those normally required are provided each way from the cutoff for a distance equal to 75 percent of the effective depth of the member. Area and spacing of the excess stirrups should be such that

$$A_v \geq 60 \frac{b_w s}{f_y}$$

where A_v = stirrup cross-sectional area, in^2 (mm^2)

b_w = web width, in (mm)

s = stirrup spacing, in (mm)

f_y = yield strength of stirrup steel, (lb/in^2) (MPa)

Stirrup spacing s should not exceed $d/8\beta_b$, where β_b is the ratio of the area of bars cut off to the total area of tension bars at the section and d is the effective depth of the member.

ULTIMATE-STRENGTH DESIGN OF RECTANGULAR BEAMS WITH COMPRESSION BARS

The bending-moment capacity of a rectangular beam with both tension and compression steel is

$$M_u = 0.90\left[(A_s - A'_s)f_y\left(d - \frac{a}{2}\right) + A'_s f_y(d - d')\right]$$

where a = depth of equivalent rectangular compressive stress distribution

$= (A_s - A'_s)f_y / f'_c b$

b = width of beam, in (mm)

d = distance from extreme compression surface to centroid of tensile steel, in (mm)

d' = distance from extreme compression surface to centroid of compressive steel, in (mm)

A_s = area of tensile steel, in^2 (mm^2)

A'_s = area of compressive steel, in^2 (mm^2)

f_y = yield strength of steel, ksi (MPa)

f'_c = 28-day strength of concrete, ksi (MPa)

This is valid only when the compressive steel reaches f_y and occurs when

$$(\rho - \rho') \geq 0.85\beta_1 \frac{f'_c d'}{f_y d} \frac{87{,}000}{87{,}000 - f_y}$$

where $\rho = A_s/bd$

$\rho' = A_s'/bd$

β_1 = a constant

WORKING-STRESS DESIGN OF RECTANGULAR BEAMS WITH COMPRESSION BARS

The following formulas, based on the linear variation of stress and strain with distance from the neutral axis, may be used in design:

$$k = \frac{1}{1 + f_s/nf_c}$$

where f_s = stress in tensile steel, ksi (MPa)

f_c = stress in extreme compression surface, ksi (MPa)

n = modular ratio, E_s/E_c

$$f_s' = \frac{kd - d'}{d - kd} 2f_s$$

where f_s' = stress in compressive steel, ksi (MPa)

d = distance from extreme compression surface to centroid of tensile steel, in (mm)

d' = distance from extreme compression surface to centroid of compressive steel, in (mm)

The factor 2 is incorporated into the preceding equation in accordance with ACI 318, "Building Code Requirements for Reinforced Concrete," to account for the effects of creep and nonlinearity of the stress–strain diagram for concrete. However, f_s' should not exceed the allowable tensile stress for the steel.

Because total compressive force equals total tensile force on a section,

$$C = C_c + C_s' = T$$

where C = total compression on beam cross section, kip (N)

C_c = total compression on concrete, kip (N) at section

C_s' = force acting on compressive steel, kip (N)

T = force acting on tensile steel, kip (N)

$$\frac{f_s}{f_c} = \frac{k}{2[\rho - \rho'(kd - d')/(d - kd)]}$$

where $\rho = A_s/bd$ and $\rho' = A_s'/bd$.

For reviewing a design, the following formulas may be used:

$$k = \sqrt{2n\left(\rho + \rho'\frac{d'}{d}\right) + n^2(\rho + \rho')^2} - n(\rho + \rho')$$

$$\bar{z} = \frac{(k^3d/3) + 4n\rho'd'[k - (d'/d)]}{k^2 + 4n\rho'[k - (d'/d)]} \qquad jd = d - \bar{z}$$

where jd is the distance between the centroid of compression and the centroid of the tensile steel. The moment resistance of the tensile steel is

$$M_s = Tjd = A_s f_s jd \qquad f_s = \frac{M}{A_s jd}$$

where M is the bending moment at the section of beam under consideration. The moment resistance in compression is

$$M_c = \frac{1}{2} f_c jbd^2 \left[k + 2n\rho' \left(1 - \frac{d'}{kd}\right)\right]$$

$$f_c = \frac{2M}{jbd^2\{k + 2n\rho'[1 - d'/kd)]\}}$$

Computer software is available for the preceding calculations. Many designers, however, prefer the following approximate formulas:

$$M_1 = \frac{1}{2} f_c bkd \left(d - \frac{kd}{3}\right)$$

$$M_s' = M - M_1 = 2f_s'A_s'(d - d')$$

where M = bending moment

M_s' = moment-resisting capacity of compressive steel

M_1 = moment-resisting capacity of concrete

ULTIMATE-STRENGTH DESIGN OF I AND T BEAMS

When the neutral axis lies in the flange, the member may be designed as a rectangular beam, with effective width b and depth d. For that condition, the flange thickness t will be greater than the distance c from the extreme compression surface to the neutral axis,

$$c = \frac{1.18\omega d}{\beta_1}$$

where β_1 = constant

$\omega = A_s f_y / bdf'_c$

A_s = area of tensile steel, in^2 (mm^2)

f_y = yield strength of steel, ksi (MPa)

f'_c = 28-day strength of concrete, ksi (MPa)

When the neutral axis lies in the web, the ultimate moment should not exceed

$$M_u = 0.90\left[(A_s - A_{sf})f_y\left(d - \frac{a}{2}\right) + A_{sf} f_y\left(d - \frac{t}{2}\right)\right] \quad (8.51)$$

where A_{sf} = area of tensile steel required to develop compressive strength of overhanging flange, in^2 (mm^2) = $0.85(b - b_w)tf'_c / f_y$

b_w = width of beam web or stem, in (mm)

a = depth of equivalent rectangular compressive stress distribution, in (mm)

$= (A_s - A_{sf})f_y / 0.85 f'_c b_w$

The quantity $\rho_w - \rho_f$ should not exceed $0.75\rho_b$, where ρ_b is the steel ratio for balanced conditions $\rho_w = A_s/b_w d$, and $\rho_f = A_{sf}/b_w d$.

WORKING-STRESS DESIGN OF I AND T BEAMS

For T beams, effective width of compression flange is determined by the same rules as for ultimate-strength design. Also, for working-stress design, two cases may occur: the neutral axis may lie in the flange or in the web. (For negative moment, a T beam should be designed as a rectangular beam with width b equal to that of the stem.)

If the neutral axis lies in the flange, a T or I beam may be designed as a rectangular beam with effective width b. If the neutral axis lies in the web or stem, an I or T beam may be designed by the following formulas, which ignore the compression in the stem, as is customary:

$$k = \frac{I}{1 + f_s/nf_c}$$

where kd = distance from extreme compression surface to neutral axis, in (mm)

d = distance from extreme compression surface to centroid of tensile steel, in (mm)

f_s = stress in tensile steel, ksi (MPa)

f_c = stress in concrete at extreme compression surface, ksi (MPa)

n = modular ratio = E_s/E_c

Because the total compressive force C equals the total tension T,

$$C = \frac{1}{2} f_c (2kd - t) \frac{bt}{kd} = T = A_s f_s$$

$$kd = \frac{2ndA_s + bt^2}{2nA_s + 2bt}$$

where A_s = area of tensile steel, in^2 (mm^2); and t = flange thickness, in (mm).

The distance between the centroid of the area in compression and the centroid of the tensile steel is

$$jd = d - \bar{z} \qquad \bar{z} = \frac{t(3kd - 2t)}{3(2kd - t)}$$

The moment resistance of the steel is

$$M_s = Tjd + A_s f_s jd$$

The moment resistance of the concrete is

$$M_c = Cjd = \frac{f_c btjd}{2kd}(2kd - t)$$

In design, M_s and M_c can be approximated by

$$M_s = A_s f_s \left(d - \frac{t}{2}\right)$$

$$M_c = \frac{1}{2} f_c bt \left(d - \frac{t}{2}\right)$$

derived by substituting $d - t/2$ for jd and $f_c/2$ for $f_c(1 - t/2kd)$, the average compressive stress on the section.

ULTIMATE-STRENGTH DESIGN FOR TORSION

When the ultimate torsion T_u is less than the value calculated from the T_u equation that follows, the area A_v of shear reinforcement should be at least

$$A_v = 50\frac{b_w s}{f_y}$$

However, when the ultimate torsion exceeds T_u calculated from the T_u equation that follows, and where web reinforcement is required, either nominally or by calculation, the minimum area of closed stirrups required is

$$A_v + 2A_t = \frac{50 b_w s}{f_y}$$

where A_t is the area of one leg of a closed stirrup resisting torsion within a distance s.

Torsion effects should be considered whenever the ultimate torsion exceeds

$$T_u = \phi\left(0.5\sqrt{f_c'}\ \Sigma x^2 y\right)$$

where ϕ = capacity reduction factor = 0.85

T_u = ultimate design torsional moment

$\Sigma x^2 y$ = sum for component rectangles of section of product of square of shorter side and longer side of each rectangle (where T section applies, overhanging flange width used in design should not exceed three times flange thickness)

The torsion T_c carried by the concrete alone should not exceed

$$T_c = \frac{0.8\sqrt{f_c'}\ \Sigma x^2 y}{\sqrt{1 + (0.4V_u/C_t T_u)^2}}$$

where $C_t = b_w d/\Sigma x^2 y$.

Spacing of closed stirrups for torsion should be computed from

$$s = \frac{A_t \phi f_y \alpha_t x_1 y_1}{(T_u - \phi T_c)}$$

where A_t = area of one leg of closed stirrup

$\alpha_t = 0.66 + 0.33y_1/x_1$ but not more than 1.50

f_y = yield strength of torsion reinforcement

x_1 = shorter dimension c to c of legs of closed stirrup

y_1 = longer dimension c to c of legs of closed stirrup

The spacing of closed stirrups, however, should not exceed $(x_1 + y_1)/4$ or 12 in (304.8 mm). Torsion reinforcement should be provided over at least a distance of $d + b$ beyond the point where it is theoretically required, where b is the beam width.

At least one longitudinal bar should be placed in each corner of the stirrups. Size of longitudinal bars should be at least No. 3, and their spacing around the perimeters of the stirrups should not exceed 12 in (304.8 mm). Longitudinal bars larger than No. 3 are required if indicated by the larger of the values of Al computed from the following two equations:

$$Al = 2A_t \frac{x_1 + y_1}{s}$$

$$Al = \left[\frac{400xs}{f_y} \left(\frac{T_u}{(T_u + V_u/3C_t)} \right) - 2A_t \right] \left(\frac{x_1 + y_1}{s} \right)$$

In the second of the preceding two equations $50b_w s/f_y$ may be substituted for $2A_t$.

The maximum allowable torsion is $T_u = \phi 5T_c$.

WORKING-STRESS DESIGN FOR TORSION

Torsion effects should be considered whenever the torsion T due to service loads exceeds

$$T = 0.55(0.5 f_c' \Sigma x^2 y)$$

where $\Sigma x^2 y$ = sum for the component rectangles of the section of the product of the square of the shorter side and the longer side of each rectangle. The allowable torsion stress on the concrete is 55 percent of that computed from the preceding T_c equation. Spacing of closed stirrups for torsion should be computed from

$$s = \frac{3A_t \alpha_t x_1 y_1 f_v}{(v_t - v_{tc})\Sigma x^2 y}$$

where A_t = area of one leg of closed stirrup

$\alpha_t = 0.66 + \dfrac{0.33y_1}{x_1}$, but not more than 1.50

v_{tc} = allowable torsion stress on concrete

x_1 = shorter dimension c to c of legs of closed stirrup

y_1 = longer dimension c to c of legs of closed stirrup

FLAT-SLAB CONSTRUCTION

Slabs supported directly on columns, without beams or girders, are classified as flat slabs. Generally, the columns flare out at the top in capitals (Fig. 5.3). However, only the portion of the inverted truncated cone thus formed that lies inside a 90° vertex angle is considered effective in resisting stress. Sometimes, the capital for an exterior column is a bracket on the inner face.

The slab may be solid, hollow, or waffle. A waffle slab usually is the most economical type for long spans, although formwork may be more expensive than for a solid slab. A waffle slab omits much of the concrete that would be in tension and thus is not considered effective in resisting stresses.

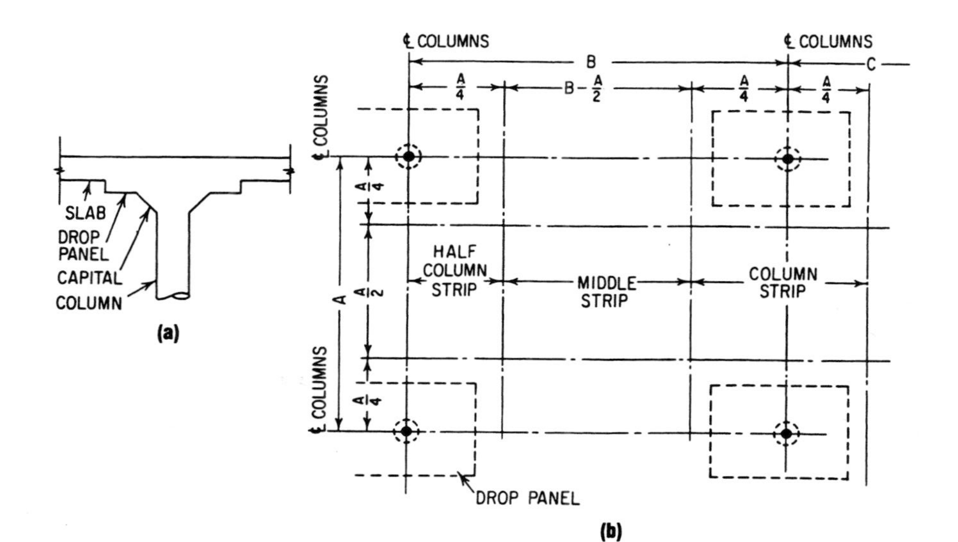

FIGURE 5.3 Concrete flat slab: (*a*) Vertical section through drop panel and column at a support. (*b*) Plan view indicates division of slab into column and middle strips.

To control deflection, the ACI *Code* establishes minimum thicknesses for slabs, as indicated by the following equation:

$$h = \frac{l_n(0.8 + f_y/200{,}000)}{36 + 5\beta[\alpha_m - 0.12(1 + 1/\beta)]}$$

$$\geq \frac{l_n(0.8 + f_y/200{,}000)}{36 + 9\beta}$$

where h = slab thickness, in (mm)

l_n = length of clear span in long direction, in (mm)

f_y = yield strength of reinforcement, ksi (MPa)

β = ratio of clear span in long direction to clear span in the short direction

α_m = average value of α for all beams on the edges of a panel

α = ratio of flexural stiffness $E_{cb}I_b$ of beam section to flexural stiffness $E_{cs}I_s$ of width of slab bounded laterally by centerline of adjacent panel, if any, on each side of beam

E_{cb} = modulus of elasticity of beam concrete

E_{cs} = modulus of elasticity of slab concrete

I_b = moment of inertia about centroidal axis of gross section of beam, including that portion of slab on each side of beam that extends a distance equal to the projection of the beam above or below the slab, whichever is greater, but not more than four times slab thickness

I_s = moment of inertia about centroidal axis of gross section of slab = $h^3/12$ times slab width specified in definition of α

Slab thickness h, however, need not be larger than $(l_n/36)(0.8 + f_y/200{,}000)$.

FLAT-PLATE CONSTRUCTION

Flat slabs with constant thickness between supports are called flat plates. Generally, capitals are omitted from the columns.

Exact analysis or design of flat slabs or flat plates is very complex. It is common practice to use approximate methods. The ACI *Code* presents two such methods: direct design and equivalent frame.

In both methods, a flat slab is considered to consist of strips parallel to column lines in two perpendicular directions. In each direction, a *column strip* spans between columns and has a width of one-fourth the shorter of the two perpendicular spans on each side of the column centerline. The portion of a slab between parallel column strips in each panel is called the *middle strip* (see Fig. 5.3).

Direct Design Method

This may be used when all the following conditions exist:

The slab has three or more bays in each direction.

Ratio of length to width of panel is 2 or less.

Loads are uniformly distributed over the panel.

Ratio of live to dead load is 3 or less.

Columns form an approximately rectangular grid (10 percent maximum offset).

Successive spans in each direction do not differ by more than one-third of the longer span.

When a panel is supported by beams on all sides, the relative stiffness of the beams satisfies

$$0.2 \leq \frac{\alpha_1}{\alpha_2}\left(\frac{l_2}{l_1}\right)^2 \leq 5$$

where $\alpha_1 = \alpha$ in direction of l_1

$\alpha_2 = \alpha$ in direction of l_2

$\alpha =$ relative beam stiffness defined in the preceding equation

$l_1 =$ span in the direction in which moments are being determined, c to c of supports

$l_2 =$ span perpendicular to l_1, c to c of supports

The basic equation used in direct design is the total static design moment in a strip bounded laterally by the centerline of the panel on each side of the centerline of the supports:

$$M_o = \frac{w l_2 l_n^2}{8}$$

where $w =$ uniform design load per unit of slab area and $l_n =$ clear span in direction moments are being determined.

The strip, with width l_2, should be designed for bending moments for which the sum in each span of the absolute values of the positive and average negative moments equals or exceeds M_o.

1. The sum of the flexural stiffnesses of the columns above and below the slab ΣK_c should be such that

$$\alpha_c = \frac{\Sigma K_c}{\Sigma(K_s + K_b)} \geq \alpha_{\min}$$

where K_c = flexural stiffness of column $= E_{cc} I_c$

E_{cc} = modulus of elasticity of column concrete

I_c = moment of inertia about centroidal axis of gross section of column

$K_s = E_{cs} I_s$

$K_b = E_{cb} I_b$

$\alpha_{\min}$ = minimum value of α_c as given in engineering handbooks

2. If the columns do not satisfy condition 1, the design positive moments in the panels should be multiplied by the coefficient:

$$\delta_s = 1 + \frac{2 - \beta_a}{4 + \beta_a}\left(1 - \frac{\alpha_c}{\alpha_{\min}}\right)$$

SHEAR IN SLABS

Slabs should also be investigated for shear, both beam type and punching shear. For beam-type shear, the slab is considered as a thin, wide rectangular beam. The critical section for diagonal tension should be taken at a distance from the face of the column or capital equal to the effective depth d of the slab. The critical section extends across the full width b of the slab. Across this section, the nominal shear stress v_u on the unreinforced concrete should not exceed the ultimate capacity $2\sqrt{f_c'}$ or the allowable working stress $1.1\sqrt{f_c'}$, where f_c' is the 28-day compressive strength of the concrete, lb/in^2 (MPa).

Punching shear may occur along several sections extending completely around the support, for example, around the face of the column or column capital or around the drop panel. These critical sections occur at a distance $d/2$ from the faces of the supports, where d is the effective depth of the slab or drop panel. Design for punching shear should be based on

$$\phi V_n = \phi(V_c + V_S)$$

where ϕ = capacity reduction factor (0.85 for shear and torsion), with shear strength V_n taken not larger than the concrete strength V_c calculated from

$$V_c = \left(2 + \frac{4}{\beta_c}\right)\sqrt{f_c'}\, b_o d \le 4\sqrt{f_c'}\, b_o d$$

where b_o = perimeter of critical section and β_c = ratio of long side to short side of critical section.

However, if shear reinforcement is provided, the allowable shear may be increased a maximum of 50 percent if shear reinforcement consisting of bars is used and increased a maximum of 75 percent if shearheads consisting of two pairs of steel shapes are used.

Shear reinforcement for slabs generally consists of bent bars and is designed in accordance with the provisions for beams with the shear strength of the concrete at critical sections taken as $2\sqrt{f_c'}\, b_o d$ at ultimate strength and $V_n \leq 6\sqrt{f_c'} b_o d$. Extreme care should be taken to ensure that shear reinforcement is accurately placed and properly anchored, especially in thin slabs.

COLUMN MOMENTS

Another important consideration in design of two-way slab systems is the transfer of moments to columns. This is generally a critical condition at edge columns, where the unbalanced slab moment is very high due to the one-sided panel.

The unbalanced slab moment is considered to be transferred to the column partly by flexure across a critical section, which is $d/2$ from the periphery of the column, and partly by eccentric shear forces acting about the centroid of the critical section.

That portion of unbalanced slab moment M_u transferred by the eccentricity of the shear is given by $\gamma_v M_u$:

$$\gamma_v = 1 - \frac{1}{1 + \left(\frac{2}{3}\right)\sqrt{\frac{b_1}{b_2}}}$$

where b_1 = width, in (mm), of critical section in the span direction for which moments are being computed; and b_2 = width, in (mm), of critical section in the span direction perpendicular to b_1.

As the width of the critical section resisting moment increases (rectangular column), that portion of the unbalanced moment transferred by flexure also increases. The maximum factored shear, which is determined by combining the vertical load and that portion of shear due to the unbalanced moment being transferred, should not exceed ϕV_c, with V_c given by preceding the V_c equation. The shear due to moment transfer can be determined at the critical section by treating this section as an analogous tube with thickness d subjected to a bending moment $\gamma_v M_u$.

The shear stress at the crack, at the face of the column or bracket support, is limited to $0.2f_c'$ or a maximum of 800 A_c, where A_c is the area of the concrete section resisting shear transfer.

The area of shear-friction reinforcement A_{vf} required in addition to reinforcement provided to take the direct tension due to temperature changes or shrinkage should be computed from

$$A_{vf} = \frac{V_u}{\phi f_y \mu}$$

where V_u is the design shear, kip (kN), at the section; f_y is the reinforcement yield strength, but not more than 60 ksi (413.7 MPa); and μ, the coefficient of friction, is 1.4 for monolithic concrete, 1.0 for concrete placed against hardened concrete, and 0.7 for concrete placed against structural rolled-steel members. The shear-friction reinforcement should be well distributed across the face of the crack and properly anchored at each side.

SPIRALS

This type of transverse reinforcement should be at least $^3/_8$ in (9.5 mm) in diameter. A spiral may be anchored at each of its ends by $1^1/_2$ extra turns of the spiral. Splices may be made by welding or by a lap of 48 bar diameters, but at least 12 in (304.8 mm). Spacing (pitch) of spirals should not exceed 3 in (76.2 mm), or be less than 1 in (25.4 mm). Clear spacing should be at least $1\frac{1}{3}$ times the maximum size of coarse aggregate.

The ratio of the volume of spiral steel/volume of concrete core (out to out of spiral) should be at least

$$\rho_s = 0.45\left(\frac{A_g}{A_c} - 1\right)\frac{f'_c}{f_y}$$

where A_g = gross area of column

A_c = core area of column measured to outside of spiral

f_y = spiral steel yield strength

f'_c = 28-day compressive strength of concrete

BRACED AND UNBRACED FRAMES

As a guide in judging whether a frame is braced or unbraced, note that the commentary on ACI 318−83 indicates that a frame may be considered braced if the bracing elements, such as shear walls, shear trusses, or other means resisting lateral movement in a story, have a total stiffness at least six times the sum of the stiffnesses of all the columns resisting lateral movement in that story.

The slenderness effect may be neglected under the two following conditions:

For columns braced against sidesway, when

$$\frac{kl_u}{r} < 34 - 12\frac{M_1}{M_2}$$

where M_1 = smaller of two end moments on column as determined by conventional elastic frame analysis, with positive sign if column is bent in single curvature and negative sign if column is bent in double curvature; and M_2 = absolute value of larger of the two end moments on column as determined by conventional elastic frame analysis.

For columns not braced against sidesway, when

$$\frac{kl_u}{r} < 22$$

LOAD-BEARING WALLS

These are subject to axial compression loads in addition to their own weight and, where there is eccentricity of load or lateral loads, to flexure. Load-bearing walls may be designed in a manner similar to that for columns but including the design requirements for non-load-bearing walls.

As an alternative, load-bearing walls may be designed by an empirical procedure given in the ACI *Code* when the eccentricity of the resulting compressive load is equal to or less than one-sixth the thickness of the wall.

Load-bearing walls designed by either method should meet the minimum reinforcing requirements for non-load-bearing walls.

In the empirical method the axial capacity, kip (kN), of the wall is

$$\phi P_n = 0.55\phi f_c' A_g \left[1 - \left(\frac{kl_c}{32h}\right)^2\right]$$

where f_c' = 28-day compressive strength of concrete, ksi (MPa)

A_g = gross area of wall section, in^2 (mm^2)

ϕ = strength reduction factor = 0.70

l_c = vertical distance between supports, in (mm)

h = overall thickness of wall, in (mm)

k = effective-length factor

For a wall supporting a concentrated load, the length of wall effective for the support of that concentrated load should be taken as the smaller of the distance center to center between loads and the bearing width plus $4h$.

SHEAR WALLS

Walls subject to horizontal shear forces in the plane of the wall should, in addition to satisfying flexural requirements, be capable of resisting the shear. The nominal shear stress can be computed from

$$v_u = \frac{V_u}{\phi hd}$$

where V_u = total design shear force

ϕ = capacity reduction factor = 0.85

$d = 0.8l_w$

h = overall thickness of wall

l_w = horizontal length of wall

The shear V_c carried by the concrete depends on whether N_u, the design axial load, lb (N), normal to the wall horizontal cross section and occurring simultaneously with V_u at the section, is a compression or tension force. When N_u is a compression force, V_c may be taken as $2\sqrt{f_c'}\,hd$, where f_c' is the 28-day strength of concrete, lb/in^2 (MPa). When N_u is a tension force, V_c should be taken as the smaller of the values calculated from

$$V_c = 3.3\sqrt{f_c'}\,hd - \frac{N_u d}{4l_w}$$

$$V_c = hd\left[0.6\sqrt{f_c'} + \frac{l_w(1.25\sqrt{f_c'} - 0.2N_u/l_w h)}{M_u/V_u - l_w/2}\right]$$

This equation does not apply, however, when $M_u/V_u - l_w/2$ is negative.

When the factored shear V_u is less than $0.5\phi V_c$, reinforcement should be provided as required by the empirical method for bearing walls.

When V_u exceeds $0.5\phi V_c$, horizontal reinforcement should be provided with $V_s = A_v f_y d/s_2$, where s_2 = spacing of horizontal reinforcement, and A_v = reinforcement area. Also, the ratio ρ_h of horizontal shear reinforcement, to the gross concrete area of the vertical section of the wall should be at least 0.0025. Spacing of horizontal shear bars should not exceed $l_w/5$, $3h$, or 18 in (457.2 mm). In addition, the ratio of vertical shear reinforcement area to gross concrete area of the horizontal section of wall does not need to be greater than that required for horizontal reinforcement but should not be less than

$$\rho_n = 0.0025 + 0.5\left(2.5 - \frac{h_w}{l_w}\right)$$

$$(\rho_h - 0.0025) \leq 0.0025$$

where h_w = total height of wall. Spacing of vertical shear reinforcement should not exceed $l_w/3$, $3h$, or 18 in (457.2 mm).

In no case should the shear strength V_n be taken greater than $10\sqrt{f_c'}\,hd$ at any section.

Bearing stress on the concrete from anchorages of post-tensioned members with adequate reinforcement in the end region should not exceed f_b calculated from

$$f_b = 0.8 f_c' \sqrt{\frac{A_b}{A_b} - 0.2} \leq 1.25 f_{ci}'$$

$$f_b = 0.6\sqrt{f_c'}\sqrt{\frac{A_b}{A_b'}} \leq f_c'$$

where A_b = bearing area of anchor plate, and A_b' = maximum area of portion of anchorage surface geometrically similar to and concentric with area of anchor plate.

A more refined analysis may be applied in the design of the end-anchorage regions of prestressed members to develop the ultimate strength of the tendons. ϕ should be taken as 0.90 for the concrete.

CONCRETE GRAVITY RETAINING WALLS

Forces acting on gravity walls include the weight of the wall, weight of the earth on the sloping back and heel, lateral earth pressure, and resultant soil pressure on the base. It is advisable to include a force at the top of the wall to account for frost action, perhaps 700 lb/linear ft (1042 kg/m). A wall, consequently, may fail by overturning or sliding, overstressing of the concrete or settlement due to crushing of the soil.

Design usually starts with selection of a trial shape and dimensions, and this configuration is checked for stability. For convenience, when the wall is of constant height, a 1-ft (0.305 m) long section may be analyzed. Moments are taken about the toe. The sum of the righting moments should be at least 1.5 times the sum of the overturning moments. To prevent sliding,

$$\mu R_v \geq 1.5P_h$$

where μ = coefficient of sliding friction

R_v = total downward force on soil, lb (N)

P_h = horizontal component of earth thrust, lb (N)

Next, the location of the vertical resultant R_v should be found at various sections of the wall by taking moments about the toe and dividing the sum by R_v. The resultant should act within the middle third of each section if there is to be no tension in the wall.

Finally, the pressure exerted by the base on the soil should be computed to ensure that the allowable pressure is not exceeded. When the resultant is within the middle third, the pressures, lb/ft^2 (Pa), under the ends of the base are given by

$$p = \frac{R_v}{A} \pm \frac{Mc}{I} = \frac{R_v}{A}\left(1 \pm \frac{6e}{L}\right)$$

where A = area of base, ft^2 (m^2)

L = width of base, ft (m)

e = distance, parallel to L, from centroid of base to R_v, ft (m)

Figure 5.4 shows the pressure distribution under a 1-ft (0.305-m) strip of wall for $e = L/2 - a$, where a is the distance of R_v from the toe. When R_v is exactly $L/3$ from the toe, the pressure at the heel becomes zero (Fig. 5.4c). When

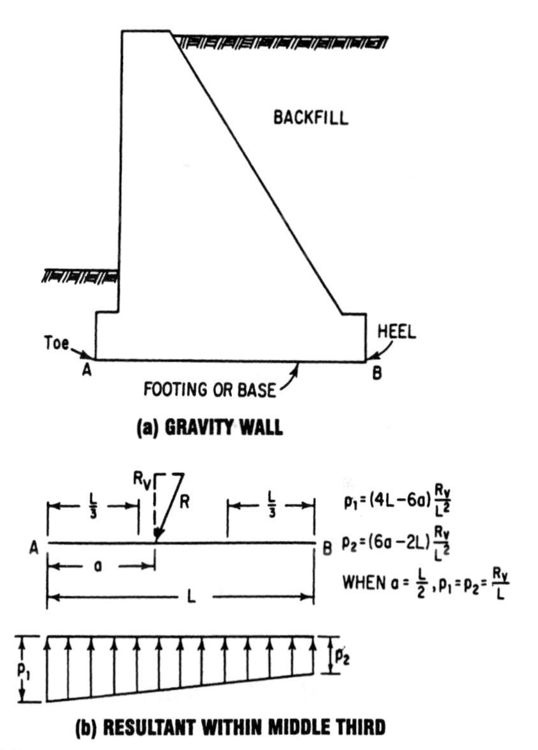

FIGURE 5.4 Diagrams for pressure of the base of a concrete gravity wall on the soil below. (a) Vertical section through the wall. (b) Significant compression under the entire base.

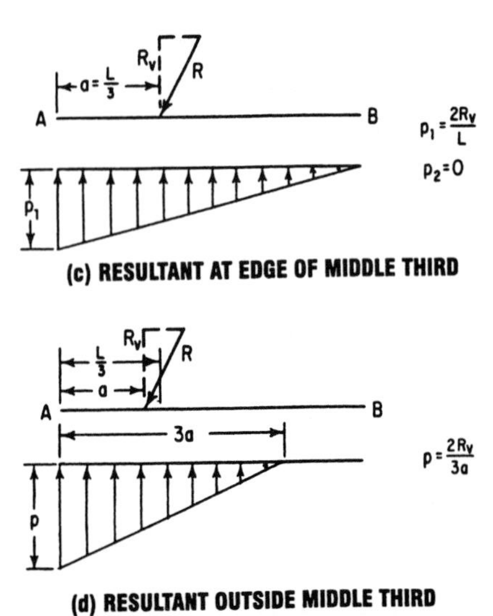

FIGURE 5.4 (*Continued*) Diagrams for pressure of the base of a concrete wall on the soil below. (*c*) No compression along one edge of the base. (*d*) Compression only under part of the base. No support from the soil under the rest of the beam.

R_v falls outside the middle third, the pressure vanishes under a zone around the heel, and pressure at the toe is much larger than for the other cases (Fig. 5.4*d*).

CANTILEVER RETAINING WALLS

This type of wall resists the lateral thrust of earth pressure through cantilever action of a vertical stem and horizontal

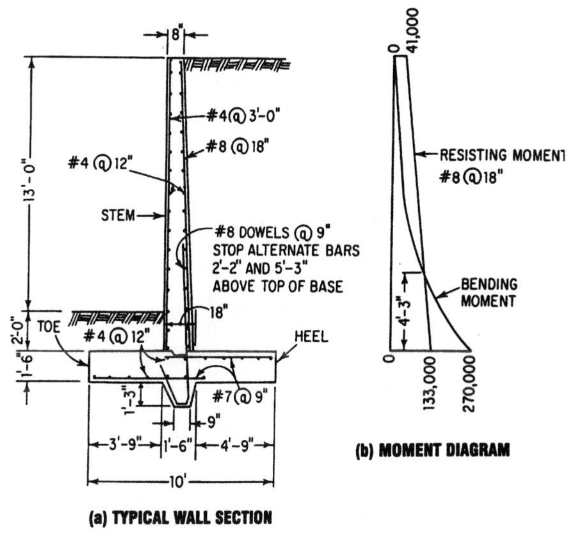

FIGURE 5.5 Cantilever retaining wall. (*a*) Vertical section shows main reinforcing steel placed vertically in the stem. (*b*) Moment diagram.

base (Fig. 5.5). Cantilever walls generally are economical for heights from 10 to 20 ft (3 to 6 m). For lower walls, gravity walls may be less costly; for taller walls, counterforts (Fig. 5.6) may be less expensive.

Shear unit stress on a horizontal section of a counterfort may be computed from $v_c = V_1/bd$, where b is the thickness of the counterfort and d is the horizontal distance from face of wall to main steel,

$$V_1 = V - \frac{M}{d}(\tan\theta + \tan\phi)$$

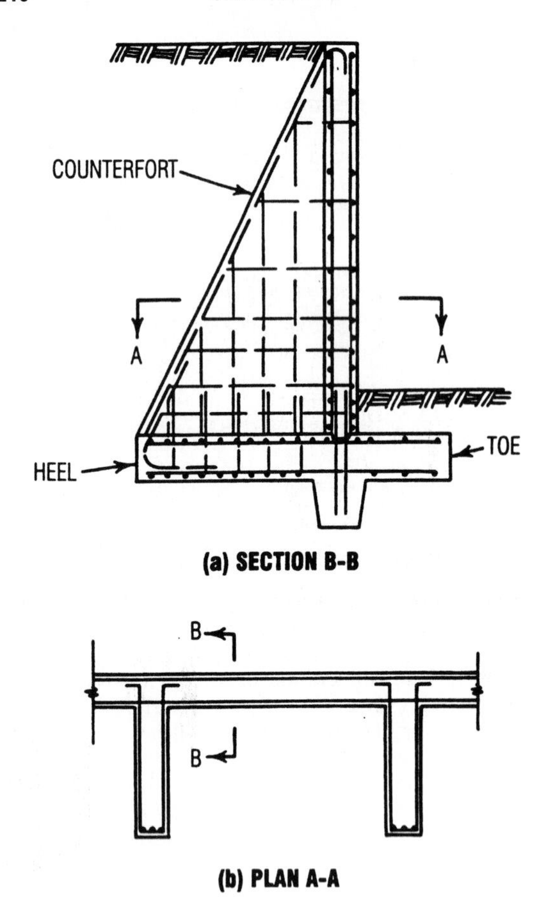

FIGURE 5.6 Counterfort retaining wall. (*a*) Vertical section. (*b*) Horizontal section.

where V = shear on section

M = bending moment at section

θ = angle earth face of counterfort makes with vertical

ϕ = angle wall face makes with vertical

For a vertical wall face, $\phi = 0$ and $V_1 = V - (M/d)\tan\theta$. The critical section for shear may be taken conservatively at a distance up from the base equal to $d' \sin\theta \cos\theta$, where d' is the depth of counterfort along the top of the base.

WALL FOOTINGS

The spread footing under a wall (Fig. 5.7) distributes the wall load horizontally to preclude excessive settlement.

The footing acts as a cantilever on opposite sides of the wall under downward wall loads and upward soil pressure. For footings supporting concrete walls, the critical section for bending moment is at the face of the wall; for footings under masonry walls, halfway between the middle and edge of the wall. Hence, for a 1-ft (0.305-m) long strip of symmetrical concrete-wall footing, symmetrically loaded, the maximum moment, ft·lb (N·m), is

$$M = \frac{p}{8}(L - a)^2$$

where p = uniform pressure on soil, lb/ft^2 (Pa)

L = width of footing, ft (m)

a = wall thickness, ft (m)

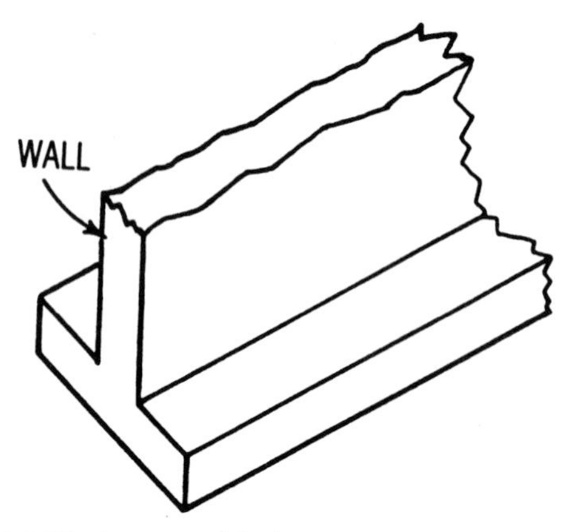

FIGURE 5.7 Concrete wall footing.

If the footing is sufficiently deep that the tensile bending stress at the bottom, $6M/t^2$, where M is the factored moment and t is the footing depth, in (mm), does not exceed $5\phi\sqrt{f_c'}$, where f_c' is the 28-day concrete strength, lb/in^2 (MPa) and $\phi = 0.90$, the footing does not need to be reinforced. If the tensile stress is larger, the footing should be designed as a 12-in (305-mm) wide rectangular, reinforced beam. Bars should be placed across the width of the footing, 3 in (76.2 mm) from the bottom. Bar development length is measured from the point at which the critical section for moment occurs. Wall footings also may be designed by ultimate-strength theory.

CHAPTER 6

TIMBER ENGINEERING FORMULAS

GRADING OF LUMBER

Stress-grade lumber consists of three classifications:

1. *Beams and stringers*. Lumber of rectangular cross section, 5 in (127 mm) or more thick and 8 in (203 mm) or more wide, graded with respect to its strength in bending when loaded on the narrow face.
2. *Joists and planks*. Lumber of rectangular cross section, 2 in (50.8 mm) to, but not including, 5 in (127 mm) thick and 4 in (102 mm) or more wide, graded with respect to its strength in bending when loaded either on the narrow face as a joist or on the wide face as a plank.
3. *Posts and timbers*. Lumber of square, or approximately square, cross section 5×5 in (127 by 127 mm), or larger, graded primarily for use as posts or columns carrying longitudinal load, but adapted for miscellaneous uses in which the strength in bending is not especially important.

Allowable unit stresses apply only for loading for which lumber is graded.

SIZE OF LUMBER

Lumber is usually designated by a nominal size. The size of unfinished lumber is the same as the nominal size, but the dimensions of dressed or finished lumber are from $^3/_8$ to $^1/_2$ in (9.5 to 12.7 mm) smaller. Properties of a few selected standard lumber sizes, along with the formulas for these properties, are shown in Table 6.1.

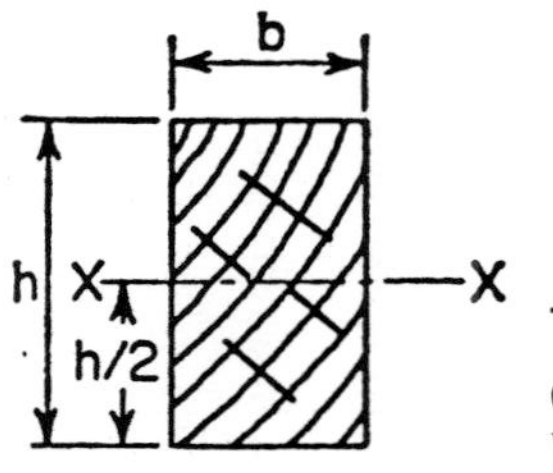

TABLE 6.1 Properties of Sections for Standard Lumber Sizes.

(Dressed (S4S) sizes, moment of inertia, and section modulus are given with respect to xx axis, with dimensions b and h, as shown on sketch)

Nominal size $b \times h$	Standard dressed size S4S $b \times h$	Area of section $A = bh^2$	Moment of inertia $I = \frac{bh^2}{12}$	Section modulus $S = \frac{bh^2}{6}$	Board feet per linear foot of piece
2×4	$1\frac{5}{8} \times 3\frac{5}{8}$	5.89	6.45	3.56	$\frac{2}{3}$
2×6	$1\frac{5}{8} \times 5\frac{1}{2}$	8.93	22.53	8.19	1
2×8	$1\frac{5}{8} \times 7\frac{1}{2}$	12.19	57.13	15.23	$1\frac{1}{3}$

Source: National Lumber Manufacturers Association.

BEARING

The allowable unit stresses given for compression perpendicular to the grain apply to bearings of any length at the ends of beams and to all bearings 6 in (152.4 mm) or more in length at other locations. When calculating the required bearing area at the ends of beams, no allowance should be made for the fact that, as the beam bends, the pressure upon the inner edge of the bearing is greater than at the end of the beam. For bearings of less than 6 in (152.4 mm) in length and not nearer than 3 in (76.2 mm) to the end of the member, the allowable stress for compression perpendicular to the grain should be modified by multiplying by the factor $(l + {}^3/_8)/l$, where l is the length of the bearing in inches (mm) measured along the grain of the wood.

BEAMS

The extreme fiber stress in bending for a rectangular timber beam is

$$f = 6M/bh^2$$
$$= M/S$$

A beam of circular cross section is assumed to have the same strength in bending as a square beam having the same cross-sectional area.

The horizontal shearing stress in a rectangular timber beam is

$$H = 3V/2bh \qquad (6.1)$$

For a rectangular timber beam with a notch in the lower face at the end, the horizontal shearing stress is

$$H = (3V/2bd_1)(h/d_1) \tag{6.2}$$

A gradual change in cross section, instead of a square notch, decreases the shearing stress nearly to that computed for the actual depth above the notch.

Nomenclature for the preceding equations follows:

f = maximum fiber stress, lb/in^2 (MPa)

M = bending moment, lb·in (Nm)

h = depth of beam, in (mm)

b = width of beam, in (mm)

S = section modulus ($= bh^2/6$ for rectangular section), in^3 (mm^3)

H = horizontal shearing stress, lb/in^2 (MPa)

V = total shear, lb (N)

d_1 = depth of beam above notch, in (mm)

l = span of beam, in (mm)

P = concentrated load, lb (N)

V_1 = modified total end shear, lb (N)

W = total uniformly distributed load, lb (N)

x = distance from reaction to concentrated load in (mm)

For simple beams, the span should be taken as the distance from face to face of supports plus one-half the required length of bearing at each end; and for continuous beams, the span should be taken as the distance between the centers of bearing on supports.

When determining V, neglect all loads within a distance from either support equal to the depth of the beam.

In the stress grade of solid-sawn beams, allowances for checks, end splits, and shakes have been made in the assigned unit stresses. For such members, Eq. (6.1) does not indicate the actual shear resistance because of the redistribution of shear stress that occurs in checked beams. For a solid-sawn beam that does not qualify using Eq. (6.1) and the H values given in published data for allowable unit stresses, the modified reaction V^1 should be determined as shown next.

For concentrated loads,

$$V^1 = \frac{10P(l - x)(x/h)^2}{9l[2 + (x/h)^2]} \tag{6.3}$$

For uniform loading,

$$V^1 = \frac{W}{2}\left(1 - \frac{2h}{l}\right) \tag{6.4}$$

The sum of the V^1 values from Eqs. (6.3) and (6.4) should be substituted for V in Eq. (6.1), and the resulting H values should be checked against those given in tables of allowable unit stresses for end-grain bearing. Such values should be adjusted for duration of loading.

COLUMNS

The allowable unit stress on timber columns consisting of a single piece of lumber or a group of pieces glued together to form a single member is

$$\frac{P}{A} = \frac{3.619E}{(l/r)^2} \tag{6.5}$$

For columns of square or rectangular cross section, this formula becomes

$$\frac{P}{A} = \frac{0.30E}{(l/d)^2} \tag{6.6}$$

For columns of circular cross section, the formula becomes

$$\frac{P}{A} = \frac{0.22E}{(l/d)^2} \tag{6.7}$$

The allowable unit stress, P/A, may not exceed the allowable compressive stress, c. The ratio, $1/d$, must not exceed 50. Values of P/A are subject to the duration of loading adjustment given previously.

Nomenclature for Eqs. (6.5) to (6.7) follows:

P = total allowable load, lb (N)

A = area of column cross section, in^2 (mm^2)

c = allowable unit stress in compression parallel to grain, lb/in^2 (MPa)

d = dimension of least side of column, in (mm)

l = unsupported length of column between points of lateral support, in (mm)

E = modulus of elasticity, lb/in^2 (MPa)

r = least radius of gyration of column, in (mm)

For members loaded as columns, the allowable unit stresses for bearing on end grain (parallel to grain) are given in data published by lumber associations. These allowable stresses apply provided there is adequate lateral support and end cuts are accurately squared and parallel. When stresses exceed 75 percent of values given, bearing must be on a snug-fitting metal plate. These stresses apply under conditions continuously dry, and must be reduced by 27 percent for glued-laminated lumber and lumber 4 in (102 mm) or less in thickness and by 9 percent for sawn lumber more than 4 in (102 mm) in thickness, for lumber exposed to weather.

COMBINED BENDING AND AXIAL LOAD

Members under combined bending and axial load should be so proportioned that the quantity

$$P_a/P + M_a/M < 1 \tag{6.8}$$

where P_a = total axial load on member, lb (N)

P = total allowable axial load, lb (N)

M_a = total bending moment on member, lb in (Nm)

M = total allowable bending moment, lb in (Nm)

COMPRESSION AT ANGLE TO GRAIN

The allowable unit compressive stress when the load is at an angle to the grain is

$$c' = c\,(c\perp)/[c\,(\sin\theta)^2 + (c\perp)\,(\cos\theta)^2] \tag{6.9}$$

where c' = allowable unit stress at angle to grain, lb/in^2 (MPa)

c = allowable unit stress parallel to grain, lb/in^2 (MPa)

$c\perp$ = allowable unit stress perpendicular to grain, lb/in^2 (MPa)

θ = angle between direction of load and direction of grain

RECOMMENDATIONS OF THE FOREST PRODUCTS LABORATORY

The *Wood Handbook* gives advice on the design of solid wood columns. (*Wood Handbook*, USDA Forest Products Laboratory, Madison, Wisc., 1999.)

Columns are divided into three categories, short, intermediate, and long. Let K denote a parameter defined by the equation

$$K = 0.64 \left(\frac{E}{f_c} \right)^{1/2} \qquad (6.10)$$

The range of the slenderness ratio and the allowable stress assigned to each category are next.

Short column,

$$\frac{L}{d} \leq 11 \qquad f = f_c \qquad (6.11)$$

Intermediate column,

$$11 < \frac{L}{d} \leq K \qquad f = f_c\left[1 - \frac{1}{3}\left(\frac{L/d}{K}\right)^4\right] \tag{6.12}$$

Long column,

$$\frac{L}{d} > K \qquad f = \frac{0.274E}{(L/d)^2} \tag{6.13}$$

The maximum L/d ratio is set at 50.

The National Design Specification covers the design of solid columns. The allowable stress in a rectangular section is as follows:

$$f = \frac{0.30E}{(L/d)^2} \qquad \text{but} \qquad f \leq f_c \tag{6.14}$$

The notational system for the preceding equations is

P = allowable load

A = sectional area

L = unbraced length

d = smaller dimension of rectangular section

E = modulus of elasticity

f_c = allowable compressive stress parallel to grain in short column of given species

f = allowable compressive stress parallel to grain in given column

COMPRESSION ON OBLIQUE PLANE

Consider that a timber member sustains a compressive force with an action line that makes an oblique angle with the grain. Let

P = allowable compressive stress parallel to grain

Q = allowable compressive stress normal to grain

N = allowable compressive stress inclined to grain

θ = angle between direction of stress N and direction of grain

By Hankinson's equation,

$$N = \frac{PQ}{P \sin_2 \theta + Q \cos^2 \theta} \qquad (6.15)$$

In Fig. 6.1, member M_1 must be notched at the joint to avoid removing an excessive area from member M_2. If the member is cut in such a manner that AC and BC make an angle of $\phi/2$ with vertical and horizontal planes, respectively, the allowable bearing pressures at these faces are identical for the two members. Let

A = sectional area of member M_1

f_1 = pressure at AC

f_2 = pressure at BC

It may be readily shown that

$$AC = b\frac{\sin(\phi/2)}{\sin\phi} \qquad BC = b\frac{\cos(\phi/2)}{\sin\phi} \qquad (6.16)$$

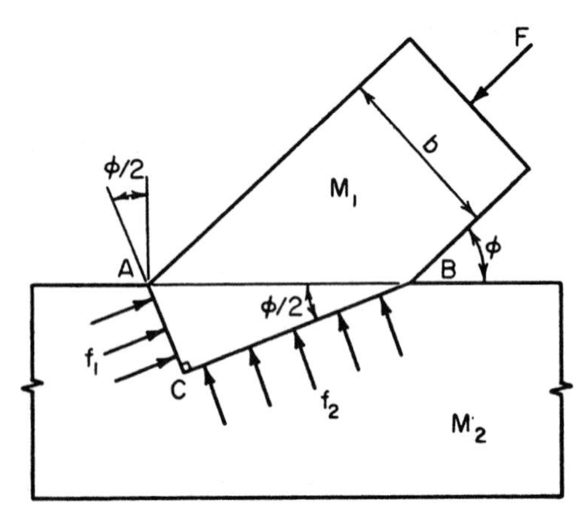

FIGURE 6.1 Timber joint.

$$f_1 = \frac{F \sin \phi}{A \tan (\phi/2)} \qquad f_2 = \frac{F \sin \phi \tan (\phi/2)}{A} \tag{6.17}$$

This type of joint is often used in wood trusses.

ADJUSTMENT FACTORS FOR DESIGN VALUES

Design values obtained by the methods described earlier should be multiplied by adjustment factors based on conditions of use, geometry, and stability. The adjustments are cumulative, unless specifically indicated in the following:

The adjusted design value F'_b for extreme-fiber bending is given by

$$F_b' = F_b C_D C_M C_t C_L C_F C_V C_{fu} C_r C_c C_f \tag{6.18}$$

where F_b = design value for extreme-fiber bending

C_D = load-duration factor

C_M = wet-service factor

C_t = temperature factor

C_L = beam stability factor

C_F = size factor—applicable only to visually graded, sawn lumber and round timber flexural members

C_v = volume factor—applicable only to glued-laminated beams

C_{fu} = flat-use factor—applicable only to dimension-lumber beams 2 to 4 in (50.8 to 101.6 mm) thick and glued-laminated beams

C_r = repetitive-member factor—applicable only to dimension-lumber beams 2 to 4 in (50.8 to 101.6 mm) thick

C_c = curvature factor—applicable only to curved portions of glued-laminated beams

C_f = form factor

For glued-laminated beams, use either C_L or C_v (whichever is smaller), not both.

The adjusted design value for tension F_t' is given by

$$F_t' = F_t C_D C_M C_t C_F \tag{6.19}$$

where F_t is the design value for tension.

For shear, the adjusted design value F_v is computed from

$$F'_V = F_V C_D C_M C_t C_H \tag{6.20}$$

where F_v is the design value for shear and C_H is the shear stress factor ≥1—permitted for F_v parallel to the grain for sawn lumber members.

For compression perpendicular to the grain, the adjusted design value $F'_{c\perp}$ is obtained from

$$F'_{c\perp} = F_{c\perp} C_M C_t C_b \tag{6.21}$$

where $F_{c\perp}$ is the design value for compression perpendicular to the grain and C_b is the bearing area factor.

For compression parallel to the grain, the adjusted design value F'_c is given by

$$F'_c = F_c C_D C_M C_t C_F C_p \tag{6.22}$$

where F_c is the design value for compression parallel to grain and C_p is the column stability factor.

For end grain in bearing parallel to the grain, the adjusted design value, F'_g is computed from

$$F'_g = F_g C_D C_t \tag{6.23}$$

where F_g is the design value for end grain in bearing parallel to the grain.

The adjusted design value for modulus of elasticity, E' is obtained from

$$E' = E C_M C_T C \cdots \tag{6.24}$$

where E = design value for modulus of elasticity

C_T = buckling stiffness factor—applicable only to sawn-lumber truss compression chords 2 × 4 in (50.8 × 101.6 mm) or smaller, when subject to combined bending and axial compression and plywood sheathing $^3/_8$ in (9.5 mm) or more thick is nailed to the narrow face

$C \cdots$ = other appropriate adjustment factors

Size and Volume Factors

For visually graded dimension lumber, design values F_b, F_t, and F_c for all species and species combinations, except southern pine, should be multiplied by the appropriate size factor C_f, given in reference data to account for the effects of member size. This factor and the factors used to develop size-specific values for southern pine are based on the adjustment equation given in *American Society for Testing and Materials* (ASTM) D1990. This equation, based on in-grade test data, accounts for differences in F_b, F_t, and F_c related to width and in F_b and F_t related to length (test span).

For visually graded timbers [5 × 5 in (127 × 127 mm) or larger], when the depth d of a stringer beam, post, or timber exceeds 12 in (304.8 mm), the design value for bending should be adjusted by the size factor

$$C_F = (12/d)^{1/9} \tag{6.25}$$

Design values for bending F_b for glued-laminated beams should be adjusted for the effects of volume by multiplying by

$$C_V = K_L\left[\left(\frac{21}{L}\right)\left(\frac{12}{d}\right)\left(\frac{5.125}{b}\right)\right]^{1/x} \tag{6.25}$$

where L = length of beam between inflection points, ft (m)

d = depth, in (mm), of beam

b = width, in (mm), of beam

= width, in (mm), of widest piece in multiple-piece layups with various widths; thus, $b \leq 10.75$ in (273 mm)

x = 20 for southern pine

= 10 for other species

K_L = loading condition coefficient

For glulam beams, the smaller of C_v and the beam stability factor C_L should be used, not both.

Radial Stresses and Curvature Factor

The radial stress induced by a bending moment in a member of constant cross section may be computed from

$$f_r = \frac{3M}{2Rbd} \tag{6.26}$$

where M = bending moment, in·lb (N·m)

R = radius of curvature at centerline of member, in (mm)

b = width of cross section, in (mm)

d = depth of cross section, in (mm)

When M is in the direction tending to decrease curvature (increase the radius), tensile stresses occur across the grain. For this condition, the allowable tensile stress across the grain is limited to one-third the allowable unit stress in horizontal shear for southern pine for all load conditions and for Douglas fir and larch for wind or earthquake loadings. The limit is 15 lb/in^2 (0.103 MPa) for Douglas fir and larch for other types of loading. These values are subject to modification for duration of load. If these values are exceeded, mechanical reinforcement sufficient to resist all radial tensile stresses is required.

When M is in the direction tending to increase curvature (decrease the radius), the stress is compressive across the grain. For this condition, the design value is limited to that for compression perpendicular to grain for all species.

For the curved portion of members, the design value for wood in bending should be modified by multiplication by the following curvature factor:

$$C_c = 1 - 2000\left(\frac{t}{R}\right)^2 \tag{6.27}$$

where t is the thickness of lamination, in (mm), and R is the radius of curvature of lamination, in (mm). Note that t/R should not exceed $^1/_{100}$ for hardwoods and southern pine or $^1/_{125}$ for softwoods other than southern pine. The curvature factor should not be applied to stress in the straight portion of an assembly, regardless of curvature elsewhere.

Bearing Area Factor

Design values for compression perpendicular to the grain $F_{c\perp}$ apply to bearing surfaces of any length at the ends of a member and to all bearings 6 in (152.4 mm) or more long at other locations. For bearings less than 6 in (152.4 mm) long and at least 3 in (76.2 mm) from the end of a member, $F_{c\perp}$ may be multiplied by the bearing area factor:

$$C_b = \frac{L_b + 0.375}{L_b} \tag{6.28}$$

where L_b is the bearing length, in (mm) measured parallel to grain. Equation (6.28) yields the values of C_b for elements with small areas, such as plates and washers, listed in reference data. For round bearing areas, such as washers, L_b should be taken as the diameter.

Column Stability and Buckling Stiffness Factors

Design values for compression parallel to the grain F_t should be multiplied by the column stability factor C_p given by Eq. (6.29):

$$C_P = \frac{1 + (F_{cE}/F_c^*)}{2c} - \sqrt{\left[\frac{1 + (F_{cE}/F_c^*)}{2c}\right]^2 - \frac{(F_{cE}/F_c^*)}{c}} \tag{6.29}$$

where F_c^* = design value for compression parallel to the grain multiplied by all applicable adjustment factors except C_p

$F_{cE} = K_{cE}E'/(L_e/d)^2$

E' = modulus of elasticity multiplied by adjustment factors

K_{cE} = 0.3 for visually graded lumber and machine-evaluated lumber

= 0.418 for products with a coefficient of variation less than 0.11

c = 0.80 for solid-sawn lumber

= 0.85 for round timber piles

= 0.90 for glued-laminated timber

For a compression member braced in all directions throughout its length to prevent lateral displacement, $C_p = 1.0$.

The buckling stiffness of a truss compression chord of sawn lumber subjected to combined flexure and axial compression under dry service conditions may be increased if the chord is 2 × 4 in (50.8 × 101.6 mm) or smaller and has the narrow face braced by nailing to plywood sheathing at least $^3/_8$ in (9.5 mm) thick in accordance with good nailing practice. The increased stiffness may be accounted for by multiplying the design value of the modulus of elasticity E by the buckling stiffness factor C_T in column stability calculations. When the effective column length L_e, in (mm), is 96 in (2.38 m) or less, C_T may be computed from

$$C_T = 1 + \frac{K_M L_e}{K_T E} \tag{6.30}$$

where K_M = 2300 for wood seasoned to a moisture content of 19 percent or less at time of sheathing attachment

= 1200 for unseasoned or partly seasoned wood at time of sheathing attachment

K_T = 0.59 for visually graded lumber and machine-evaluated lumber

= 0.82 for products with a coefficient of variation of 0.11 or less

When L_e is more than 96 in (2.38 m), C_T should be calculated from Eq. (6.30) with L_e = 96 in (2.38 m). For additional information on wood trusses with metal-plate connections, see design standards of the Truss Plate Institute, Madison, Wisconsin.

The slenderness ratio R_B for beams is defined by

$$R_B = \sqrt{\frac{L_e d}{b^2}} \tag{6.31}$$

The slenderness ratio should not exceed 50.

The effective length L_e for Eq. (6.31) is given in terms of unsupported length of beam in reference data. Unsupported length is the distance between supports or the length of a cantilever when the beam is laterally braced at the supports to prevent rotation and adequate bracing is not installed elsewhere in the span. When both rotational and lateral displacements are also prevented at intermediate points, the unsupported length may be taken as the distance between points of lateral support. If the compression edge is supported throughout the length of the beam and adequate bracing is installed at the supports, the unsupported length is zero.

The beam stability factor C_L may be calculated from

$$C_L = \frac{1 + (F_{\text{bE}}/F_b^*)}{1.9} - \sqrt{\left[\frac{1 + (F_{\text{bE}}/F_b^*)}{1.9}\right]^2 - \frac{F_{\text{bE}}/F_b^*}{0.95}} \tag{6.32}$$

where F_b^* = design value for bending multiplied by all applicable adjustment factors, except C_{fu}, C_V, and C_L

$F_{bE} = K_{bE} E'/R_B^2$

K_{bE} = 0.438 for visually graded lumber and machine-evaluated lumber

= 0.609 for products with a coefficient of variation of 0.11 or less

E' = design modulus of elasticity multiplied by applicable adjustment factors

FASTENERS FOR WOOD

Nails and Spikes

The allowable withdrawal load per inch (25.4 mm) of penetration of a common nail or spike driven into side grain (perpendicular to fibers) of seasoned wood, or unseasoned wood that remains wet, is

$$p = 1{,}380G^{5/2}D \tag{6.33}$$

where p = allowable load per inch (mm) of penetration into member receiving point, lb (N)

D = diameter of nail or spike, in (mm)

G = specific gravity of wood, oven dry

The total allowable lateral load for a nail or spike driven into side grain of seasoned wood is

$$p = CD^{3/2} \tag{6.34}$$

where p = allowable load per nail or spike, lb (N)

D = diameter of nail or spike, in (mm)

C = coefficient dependent on group number of wood (see Table 6.1)

Values of C for the four groups into which stress-grade lumber is classified are

Group I: $C = 2{,}040$
Group II: $C = 1{,}650$
Group III: $C = 1{,}350$
Group IV: $C = 1{,}080$

The loads apply where the nail or spike penetrates into the member, receiving its point at least 10 diameters for Group I species, 11 diameters for Group II species, 13 diameters for Group III species, and 14 diameters for Group IV species. Allowable loads for lesser penetrations are directly proportional to the penetration, but the penetration must be at least one-third that specified.

Wood Screws

The allowable withdrawal load per inch (mm) of penetration of the threaded portion of a wood screw into side grain of seasoned wood that remains dry is

$$p = 2{,}850G^2D \quad (6.35)$$

where p = allowable load per inch (mm) of penetration of threaded portion into member receiving point, lb (N)

D = diameter of wood screw, in (mm)

G = specific gravity of wood, oven dry (see Table 6.1)

Wood screws should not be loaded in withdrawal from end grain.

The total allowable lateral load for wood screws driven into the side grain of seasoned wood which remains dry is

$$p = CD^2 \quad (6.36)$$

where p = allowable load per wood screw, lb (N)

D = diameter of wood screw, in (mm)

C = coefficient dependent on group number of wood (Table 6.2)

TABLE 6.2 Specific Gravity and Group Number for Common Species of Lumber

Species	Group number	Specific gravity, G	G^2	$G^{5/2}$
Douglas fir	II	0.51	0.260	0.186
Pine, southern	II	0.59	0.348	0.267
Hemlock, western	III	0.44	0.194	0.128
Hemlock, eastern	IV	0.43	0.185	0.121
Pine, Norway	III	0.47	0.221	0.151
Redwood	III	0.42	0.176	0.114
Spruce	IV	0.41	0.168	0.108

Values of C for the four groups into which stress-grade lumber is classified are

Group I: $C = 4{,}800$

Group II: $C = 3{,}960$

Group III: $C = 3{,}240$

Group IV: $C = 2{,}520$

The allowable lateral load for wood screws driven into end grain is two-thirds that given for side grain.

ADJUSTMENT OF DESIGN VALUES FOR CONNECTIONS WITH FASTENERS

Nominal design values for connections or wood members with fasteners should be multiplied by applicable adjustment factors available from lumber associations and in civil engineering handbooks to obtain adjusted design values. The types of loading on the fasteners may be divided into four classes: lateral loading, withdrawal, loading parallel to grain, and loading perpendicular to grain. Adjusted design values are given in terms of nominal design values and adjustment factors in the equations below. The following variables are used in the equations:

Z' = adjusted design value for lateral loading

Z = nominal design value for lateral loading

W' = adjusted design value for withdrawal

W = nominal design value for withdrawal

P' = adjusted value for loading parallel to grain

P = nominal value for loading parallel to grain

Q' = adjusted value for loading normal to grain

Q = nominal value for loading normal to grain

For bolts,

$$Z' = ZC_DC_MC_tC_gD_\Delta$$

where C_D = load-duration factor, not to exceed 1.6 for connections

C_M = wet-service factor, not applicable to toenails loaded in withdrawal

C_t = temperature factor

C_g = group-action factor

C_Δ = geometry factor

For split-ring and shear-plate connectors,

$$P' = PC_DC_MC_tC_gC_\Delta C_dC_{st}$$

$$Q' = QC_DC_MC_tC_gC_\Delta C_d$$

where C_d is the penetration-depth factor and C_{st} is the metal-side-plate factor.

For nails and spikes,

$$W' = WC_D\,C_M\,C_t\,C_{tn}$$

$$Z' = ZC_D\,C_M\,C_t\,C_d\,C_{eg}\,C_{di}\,C_{tn}$$

where C_{di} = is the diaphragm factor and C_{tn} = toenail factor.

For wood screws,

$$W' = WC_DC_MC_t$$

$$Z' = ZC_DC_MC_tC_dC_{eg}$$

where C_{eg} is the end-grain factor.

For lag screws,

$$W' = WC_DC_MC_tC_{eg}$$

$$Z' = ZC_DC_MC_tC_gC_\Delta C_dC_{eg}$$

For metal plate connectors,

$$Z' = ZC_DC_MC_t$$

For drift bolts and drift pins,

$$W' = WC_DC_MC_tC_{eg}$$

$$Z' = ZC_DC_MC_tC_gC_\Delta C_dC_{eg}$$

For spike grids,

$$Z' = ZC_DC_MC_tC_\Delta$$

ROOF SLOPE TO PREVENT PONDING

Roof beams should have a continuous upward slope equivalent to $^1/_4$ in/ft (20.8 mm/m) between a drain and the high point of a roof, in addition to minimum recommended camber to avoid ponding. When flat roofs have insufficient slope for drainage (less than $^1/_4$ in/ft) (20.8 mm/m) the stiffness of supporting members should be such that a 5-lb/ft^2 (239.4 N/mm^2) load causes no more than $^1\!/_2$-in (12.7-mm) deflection.

Because of ponding, snow loads or water trapped by gravel stops, parapet walls, or ice dams magnify stresses and deflections from existing roof loads by

$$C_p = \frac{1}{1 - W'L^3/\pi^4 EI}$$

where C_p = factor for multiplying stresses and deflections under existing loads to determine stresses and deflections under existing loads plus ponding

W = weight of 1 in (25.4 mm) of water on roof area supported by beam, lb (N)

L = span of beam, in (mm)

E = modulus of elasticity of beam material, lb/in^2 (MPa)

I = moment of inertia of beam, in^4 (mm^4)

(Kuenzi and Bohannan, "Increases in Deflection and Stresses Caused by Ponding of Water on Roofs," Forest Products Laboratory, Madison, Wisconsin.)

BENDING AND AXIAL TENSION

Members subjected to combined bending and axial tension should be proportioned to satisfy the interaction equations

$$\frac{f_t}{F_c'} + \frac{f_b}{F_b^*} \le 1$$

and

$$\frac{(f_b - f_t)}{F_b^{**}} \leq 1$$

where f_t = tensile stress due to axial tension acting alone

f_b = bending stress due to bending moment alone

F_t' = design value for tension multiplied by applicable adjustment factors

F_b^* = design value for bending multiplied by applicable adjustment factors except C_L

F_b^{**} = design value for bending multiplied by applicable adjustment factors except C_v

The load duration factor C_D associated with the load of shortest duration in a combination of loads with differing duration may be used to calculate F_t' and F_b^*. All applicable load combinations should be evaluated to determine the critical load combination.

BENDING AND AXIAL COMPRESSION

Members subjected to a combination of bending and axial compression (beam columns) should be proportioned to satisfy the interaction equation

$$\left(\frac{f_c}{F_c'}\right)^2 + \frac{f_{b1}}{[1 - (f_c/F_{cE1})]F_{b1}'} + \frac{f_{b2}}{[1 - (f_c/F_{cE2}) - (f_{b1}/F_{bE})^2]F_{b2}'} \leq 1$$

where f_c = compressive stress due to axial compression acting alone

F'_c = design value for compression parallel to grain multiplied by applicable adjustment factors, including the column stability factor

f_{b1} = bending stress for load applied to the narrow face of the member

f_{b2} = bending stress for load applied to the wide face of the member

F'_{b1} = design value for bending for load applied to the narrow face of the member multiplied by applicable adjustment factors, including the column stability factor

F'_{b2} = design value for bending for load applied to the wide face of the member multiplied by applicable adjustment factors, including the column stability factor

For either uniaxial or biaxial bending, f_c should not exceed

$$F_{\mathrm{cE1}} = \frac{K_{\mathrm{cE}}\,E'}{(L_{e1}/d_1)^2}$$

where E' is the modulus of elasticity multiplied by adjustment factors. Also, for biaxial bending, f_c should not exceed

$$F_{\mathrm{cE2}} = \frac{K_{\mathrm{cE}}\,E'}{(L_{e2}/d_2)^2}$$

and f_{b1} should not be more than

$$F_{\mathrm{bE}} = \frac{K_{\mathrm{bE}}\,E'}{R_B^2}$$

where d_1 is the width of the wide face and d_2 is the width of the narrow face. Slenderness ratio R_B for beams is given earlier in this section. K_{bE} is defined earlier in this section. The effective column lengths L_{e1} for buckling in the d_1 direction and L_{e2} for buckling in the d_2 direction, E', F_{cE1}, and F_{cE2} should be determined as shown earlier.

As for the case of combined bending and axial tension, F'_c, F'_{b1}, and F'_{b2} should be adjusted for duration of load by applying C_D.

CHAPTER 7

SURVEYING FORMULAS

UNITS OF MEASUREMENT

Units of measurement used in past and present surveys are

For construction work: feet, inches, fractions of inches (m, mm)

For most surveys: feet, tenths, hundredths, thousandths (m, mm)

For *National Geodetic Survey* (NGS) control surveys: meters, 0.1, 0.01, 0.001 m

The most-used equivalents are

1 meter = 39.37 in (exactly) = 3.2808 ft

1 rod = 1 pole = 1 perch = $16\frac{1}{2}$ ft (5.029 m)

1 engineer's chain = 100 ft = 100 links (30.48 m)

1 Gunter's chain = 66 ft (20.11 m) = 100 Gunter's links (lk) = 4 rods = $\frac{1}{80}$ mi (0.020 km)

1 acre = 100,000 sq (Gunter's) links = 43,560 ft^2 = 160 $rods^2$ = 10 sq (Gunter's) chains = 4046.87 m^2 = 0.4047 ha

1 rood = $\frac{3}{4}$ acre (1011.5 m^2) = 40 $rods^2$ (also local unit = $5\frac{1}{2}$ to 8 yd) (5.029 to 7.315 m)

1 ha = 10,000 m^2 = 107,639.10 ft^2 = 2.471 acres

1 arpent = about 0.85 acre, or length of side of 1 square arpent (varies) (about 3439.1 m^2)

1 statute mi = 5280 ft = 1609.35 m

1 mi^2 = 640 acres (258.94 ha)

1 nautical mi (U.S.) = 6080.27 ft = 1853.248 m

1 fathom = 6 ft (1.829 m)

1 cubit = 18 in (0.457 m)

1 vara = 33 in (0.838 m) (Calif.), $33\frac{1}{3}$ in (0.851 m) (Texas), varies

1 degree = $\frac{1}{360}$ circle = 60 min = 3600 s = 0.01745 rad

sin 1° = 0.01745241

1 rad = 57° 17′ 44.8″ or about 57.30°

1 grad (grade) = $\frac{1}{400}$ circle = $\frac{1}{100}$ quadrant = 100 centesimal min = 10^4 centesimals (French)

1 mil = $\frac{1}{6400}$ circle = 0.05625°

1 military pace (milpace) = $2\frac{1}{2}$ ft (0.762 m)

THEORY OF ERRORS

When a number of surveying measurements of the same quantity have been made, they must be analyzed on the basis of probability and the theory of errors. After all systematic (cumulative) errors and mistakes have been eliminated, random (compensating) errors are investigated to determine the most probable value (*mean*) and other critical values. Formulas determined from statistical theory and the normal, or Gaussian, bell-shaped probability distribution curve, for the most common of these values follow.

Standard deviation of a series of observations is

$$\sigma_s = \pm\sqrt{\frac{\Sigma d^2}{n-1}}$$

where d = residual (difference from mean) of single observation and n = number of observations.

The *probable error* of a single observation is

$$PE_s = \pm 0.6745\sigma_s$$

(The probability that an error within this range will occur is 0.50.)

The probability that an error will lie between two values is given by the ratio of the area of the probability curve included between the values to the total area. Inasmuch as the area under the entire probability curve is unity, there is a 100 percent probability that all measurements will lie within the range of the curve.

The area of the curve between $\pm\sigma_s$ is 0.683; that is, there is a 68.3 percent probability of an error between $\pm\sigma_s$ in a single measurement. This error range is also called the one-sigma or 68.3 percent confidence level. The area of the curve between $\pm 2\sigma_s$ is 0.955. Thus there is a 95.5 percent probability of an error between $\pm 2\sigma_s$ and $\pm 2\sigma_s$ that represents the 95.5 percent error (two-sigma or 95.5 percent confidence level). Similarly, $\pm 3\sigma_s$ is referred to as the 99.7 percent error (three-sigma or 99.7 percent confidence level). For practical purposes, a maximum tolerable level often is assumed to be the 99.9 percent error. Table 7.1 indicates the probability of occurrence of larger errors in a single measurement.

The probable error of the combined effects of accidental errors from different causes is

TABLE 1 Probability of Error in a Single Measurement

Error	Confidence level, %	Probability of larger error
Probable ($0.6745\sigma_s$)	50	1 in 2
Standard deviation (σ_s)	68.3	1 in 3
90% ($1.6449\sigma_s$)	90	1 in 10
$2\sigma_s$ or 95.5%	95.5	1 in 20
$3\sigma_s$ or 97.7%	99.7	1 in 370
Maximum ($3.29\sigma_s$)	99.9+	1 in 1000

$$E_{\text{sum}} = \sqrt{E_1^2 + E_2^2 + E_3^2 + \cdots}$$

where E_1, E_2, E_3 . . . are probable errors of the separate measurements.

Error of the mean is

$$E_m = \frac{E_{\text{sum}}}{n} = \frac{E_s\sqrt{n}}{n} = \frac{E_s}{\sqrt{n}}$$

where E_s = specified error of a single measurement.

Probable error of the mean is

$$PE_m = \frac{PE_s}{\sqrt{n}} = \pm 0.6745\sqrt{\frac{\Sigma d^2}{n(n-1)}}$$

MEASUREMENT OF DISTANCE WITH TAPES

Reasonable precisions for different methods of measuring distances are

Pacing (ordinary terrain): $^1/_{50}$ to $^1/_{100}$

Taping (ordinary steel tape): $^1/_{1000}$ to $^1/_{10,000}$ (Results can be improved by use of tension apparatus, transit alignment, leveling.)

Baseline (invar tape): $^1/_{50,000}$ to $^1/_{1,000,000}$

Stadia: $^1/_{300}$ to $^1/_{500}$ (with special procedures)

Subtense bar: $^1/_{1000}$ to $^1/_{7000}$ (for short distances, with a 1-s theodolite, averaging angles taken at both ends)

Electronic distance measurement (EDM) devices have been in use since the middle of the twentieth century and have now largely replaced steel tape measurements on large projects. The continued development, and the resulting drop in prices, are making their use widespread. A knowledge of steel-taping errors and corrections remains important, however, because use of earlier survey data requires a knowledge of how the measurements were made, common sources for errors, and corrections that were typically required.

For ordinary taping, a tape accurate to 0.01 ft (0.00305 m) should be used. The tension of the tape should be about 15 lb (66.7 N). The temperature should be determined within 10°F (5.56°C); and the slope of the ground, within 2 percent; and the proper corrections, applied. The correction to be applied for temperature when using a steel tape is

$$C_t = 0.0000065s(T - T_0)$$

The correction to be made to measurements on a slope is

$$C_h = s\,(1 - \cos\theta) \qquad \text{exact}$$

or $$= 0.00015 s\theta^2 \qquad \text{approximate}$$

or $$= h^2/2s \qquad \text{approximate}$$

where C_t = temperature correction to measured length, ft (m)

C_h = correction to be subtracted from slope distance, ft (m)

s = measured length, ft (m)

T = temperature at which measurements are made, °F (°C)

T_0 = temperature at which tape is standardized, °F (°C)

h = difference in elevation at ends of measured length, ft (m)

θ = slope angle, degree

In more accurate taping, using a tape standardized when fully supported throughout, corrections should also be made for tension and for support conditions. The correction for tension is

$$C_p = \frac{(P_m - P_s)s}{SE}$$

The correction for sag when not fully supported is

$$C_s = \frac{w^2 L^3}{24 P_m^2}$$

where C_p = tension correction to measured length, ft (m)

C_s = sag correction to measured length for each section of unsupported tape, ft (m)

P_m = actual tension, lb (N)

P_s = tension at which tape is standardized, lb (N) (usually 10 lb) (44.4 N)

S = cross-sectional area of tape, in^2 (mm^2)

E = modulus of elasticity of tape, lb/in^2 (MPa) (29 million lb/in^2 (MPa) for steel) (199,955 MPa)

w = weight of tape, lb/ft (kg/m)

L = unsupported length, ft (m)

Slope Corrections

In slope measurements, the horizontal distance $H = L \cos x$, where L = slope distance and x = vertical angle, measured from the horizontal—a simple hand calculator operation. For slopes of 10 percent or less, the correction to be applied to L for a difference d in elevation between tape ends, or for a horizontal offset d between tape ends, may be computed from

$$C_s = \frac{d^2}{2L}$$

For a slope greater than 10 percent, C_s may be determined from

$$C_s = \frac{d^2}{2L} + \frac{d^4}{8L^3}$$

Temperature Corrections

For *incorrect tape length*:

$$C_t = \frac{(\text{actual tape length} - \text{nominal tape length})L}{\text{nominal tape length}}$$

For *nonstandard tension*:

$$C_t = \frac{(\text{applied pull} - \text{standard tension})L}{AE}$$

where A = cross-sectional area of tape, in^2 (mm^2); and E = modulus of elasticity = 29,000,00 lb/in^2 for steel (199,955 MPa).

For *sag correction* between points of support, ft (m):

$$C = -\frac{w^2 L_s^3}{24P^2}$$

where w = weight of tape per foot, lb (N)

L_s = unsupported length of tape, ft (m)

P = pull on tape, lb (N)

Orthometric Correction

This is a correction applied to preliminary elevations due to flattening of the earth in the polar direction. Its value is a function of the latitude and elevation of the level circuit.

Curvature of the earth causes a horizontal line to depart from a level surface. The departure C_f, ft, or C_m, (m), may be computed from

$$C_f = 0.667M^2 = 0.0239F^2$$

$$C_m = 0.0785K^2$$

where M, F, and K are distances in miles, thousands of feet, and kilometers, respectively, from the point of tangency to the earth.

Refraction causes light rays that pass through the earth's atmosphere to bend toward the earth's surface. For horizontal sights, the average angular displacement (like the sun's diameter) is about 32 min. The displacement R_f, ft, or R_m, m, is given approximately by

$$R_f = 0.093M^2 = 0.0033F^2$$

$$R_m = 0.011K^2$$

To obtain the combined effect of refraction and curvature of the earth, subtract R_f from C_f or R_m from C_m.

Borrow-pit or cross-section leveling produces elevations at the corners of squares or rectangles with sides that are dependent on the area to be covered, type of terrain, and accuracy desired. For example, sides may be 10, 20, 40, 50, or 100 ft (3.048, 6.09, 12.19, 15.24, or 30.48 m). Contours can be located readily, but topographic features, not so well. Quantities of material to be excavated or filled are computed, in yd^3 (m^3), by selecting a grade elevation or final ground elevation, computing elevation differences for the corners, and substituting in

$$Q = \frac{nxA}{108}$$

where n = number of times a particular corner enters as part of a division block

x = difference in ground and grade elevation for each corner, ft (m)

A = area of each block, ft^2 (m^2)

VERTICAL CONTROL

The NGS provides vertical control for all types of surveys. NGS furnishes descriptions and elevations of bench marks on request. As given in "Standards and Specifications for Geodetic Control Networks," Federal Geodetic Control Committee, the relative accuracy C, mm, required between directly connected bench marks for the three orders of leveling is

First order: $C = 0.5\sqrt{K}$ for Class I and $0.7\sqrt{K}$ for Class II

Second order: $C = 1.0\sqrt{K}$ for Class I and $1.3\sqrt{K}$ for Class II

Third order: $C = 2.0\sqrt{K}$

where K is the distance between bench marks, km.

STADIA SURVEYING

In stadia surveying, a transit having horizontal stadia crosshairs above and below the central horizontal crosshair is used. The difference in the rod readings at the stadia crosshairs is termed the rod intercept. The intercept may be converted to the horizontal and vertical distances between the instrument and the rod by the following formulas:

$$H = Ki(\cos a)^2 + (f + c)\cos a$$

$$V = \frac{1}{2}Ki(\sin 2a) + (f + c)\sin a$$

where H = horizontal distance between center of transit and rod, ft (m)

V = vertical distance between center of transit and point on rod intersected by middle horizontal crosshair, ft (m)

K = stadia factor (usually 100)

i = rod intercept, ft (m)

a = vertical inclination of line of sight, measured from the horizontal, degree

$f + c$ = instrument constant, ft (m) (usually taken as 1 ft) (0.3048 m)

In the use of these formulas, distances are usually calculated to the foot (meter) and differences in elevation to tenths of a foot (meter).

Figure 7.1 shows stadia relationships for a horizontal sight with the older type of external-focusing telescope. Relationships are comparable for the internal-focusing type.

For horizontal sights, the stadia distance, ft, (m) (from instrument spindle to rod), is

$$D = R\frac{f}{i} + C$$

where R = intercept on rod between two sighting wires, ft (m)

f = focal length of telescope, ft (m) (constant for specific instrument)

i = distance between stadia wires, ft (m)

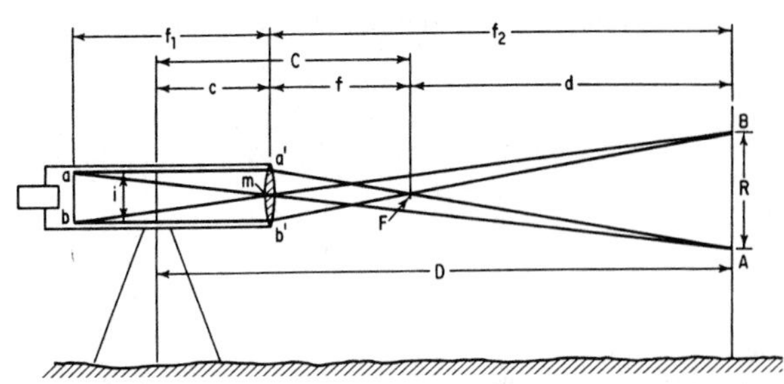

FIGURE 7.1 Distance D is measured with an external-focusing telescope by determining interval R intercepted on a rod AB by two horizontal sighting wires a and b.

$$C = f + c$$

c = distance from center of spindle to center of objective lens, ft (m)

C is called the stadia constant, although c and C vary slightly.

The value of f/i, the stadia factor, is set by the manufacturer to be about 100, but it is not necessarily 100.00. The value should be checked before use on important work, or when the wires or reticle are damaged and replaced.

PHOTOGRAMMETRY

Photogrammetry is the art and science of obtaining reliable measurements by photography (metric photogrammetry) and qualitative evaluation of image data (photo interpretation). It includes use of terrestrial, close-range, aerial, vertical, oblique, strip, and space photographs along with their interpretation.

Scale formulas are as follows:

$$\frac{\text{Photo scale}}{\text{Map scale}} = \frac{\text{photo distance}}{\text{map distance}}$$

$$\text{Photo scale} = \frac{ab}{AB} = \frac{f}{H - h_1}$$

where f = focal length of lens, in (m)

H = flying height of airplane above datum (usually mean sea level), ft (m)

h_1 = elevation of point, line, or area with respect to datum, ft (m)

CHAPTER 8

SOIL AND EARTHWORK FORMULAS

PHYSICAL PROPERTIES OF SOILS

Basic soil properties and parameters can be subdivided into physical, index, and engineering categories. Physical soil properties include density, particle size and distribution, specific gravity, and water content.

The *water content* w of a soil sample represents the weight of free water contained in the sample expressed as a percentage of its dry weight.

The *degree of saturation* S of the sample is the ratio, expressed as percentage, of the volume of free water contained in a sample to its total volume of voids V_v.

Porosity n, which is a measure of the relative amount of voids, is the ratio of void volume to the total volume V of soil:

$$n = \frac{V_v}{V} \tag{8.1}$$

The ratio of V_v to the volume occupied by the soil particles V_s defines the *void ratio* e. Given e, the degree of saturation may be computed from

$$S = \frac{wG_s}{e} \tag{8.2}$$

where G_s represents the specific gravity of the soil particles. For most inorganic soils, G_s is usually in the range of 2.67 ± 0.05.

The dry unit weight γ_d of a soil specimen with any degree of saturation may be calculated from

$$\gamma_d = \frac{\gamma_w G_s S}{1 + wG_s} \tag{8.3}$$

where γ_w is the unit weight of water and is usually taken as 62.4 lb/ft^3 (1001 kg/m^3) for freshwater and 64.0 lb/ft^3 (1026.7 kg/m^3) for seawater.

INDEX PARAMETERS FOR SOILS

Index parameters of cohesive soils include liquid limit, plastic limit, shrinkage limit, and activity. Such parameters are useful for classifying cohesive soils and providing correlations with engineering soil properties.

The *liquid limit* of cohesive soils represents a near-liquid state, that is, an undrained shear strength about 0.01 lb/ft^2 (0.0488 kg/m^2). The water content at which the soil ceases to exhibit plastic behavior is termed the *plastic limit.* The *shrinkage limit* represents the water content at which no further volume change occurs with a reduction in water content. The most useful classification and correlation parameters are the plasticity index I_p, the liquidity index I_l, the shrinkage index I_s, and the activity A_c. These parameters are defined in Table 8.1.

Relative density D_r of cohesionless soils may be expressed in terms of void ratio e or unit dry weight γ_d:

$$D_r = \frac{e_{\max} - e_o}{e_{\max} - e_{\min}} \tag{8.4}$$

$$D_r = \frac{1/\gamma_{\min} - 1/\gamma_d}{1/\gamma_{\min} - 1/\gamma_{\max}} \tag{8.5}$$

D_r provides cohesionless soil property and parameter correlations, including friction angle, permeability, compressibility, small-strain shear modulus, cyclic shear strength, and so on.

TABLE 8.1 Soil Indices

Index	Definition[†]	Correlation
Plasticity	$I_p = W_l - W_p$	Strength, compressibility, compactibility, and so forth
Liquidity	$I_l = \frac{W_n - W_p}{I_p}$	Compressibility and stress rate
Shrinkage	$I_s = W_p - W_s$	Shrinkage potential
Activity	$A_c = \frac{I_p}{\mu}$	Swell potential, and so forth

[†] W_t = liquid limit; W_p = plastic limit; W_n = moisture content, %; W_s = shrinkage limit; μ = percent of soil finer than 0.002 mm (clay size).

RELATIONSHIP OF WEIGHTS AND VOLUMES IN SOILS

The unit weight of soil varies, depending on the amount of water contained in the soil. Three unit weights are in general use: the saturated unit weight γ_{sat}, the dry unit weight γ_{dry}, and the buoyant unit weight γ_b:

$$\gamma_{sat} = \frac{(G + e)\gamma_0}{1 + e} = \frac{(1 + w)G\gamma_0}{1 + e} \qquad S = 100\%$$

$$\gamma_{dry} = \frac{G\gamma_0}{(1 + e)} \qquad S = 0\%$$

$$\gamma_b = \frac{(G - 1)\gamma_0}{1 + e} \qquad S = 100\%$$

Unit weights are generally expressed in pound per cubic foot or gram per cubic centimeter. Representative values of unit weights for a soil with a specific gravity of 2.73 and a void ratio of 0.80 are

$$\gamma_{sat} = 122 \text{ lb/ft}^3 = 1.96 \text{ g/cm}^3$$

$$\gamma_{dry} = 95 \text{ lb/ft}^3 = 1.52 \text{ g/cm}^3$$

$$\gamma_b = 60 \text{ lb/ft}^3 = 0.96 \text{ g/cm}^3$$

The symbols used in the three preceding equations and in Fig. 8.1 are

G = specific gravity of soil solids (specific gravity of quartz is 2.67; for majority of soils specific gravity ranges between 2.65 and 2.85; organic soils would have lower specific gravities)

γ_0 = unit weight of water, 62.4 lb/ft^3 (1.0 g/cm^3)

e = voids ratio, volume of voids in mass of soil divided by volume of solids in same mass; also equal to $n/(1 - n)$, where n is porosity—volume of voids in mass of soil divided by total volume of same mass

S = degree of saturation, volume of water in mass of soil divided by volume of voids in same mass

w = water content, weight of water in mass of soil divided by weight of solids in same mass; also equal to Se/G

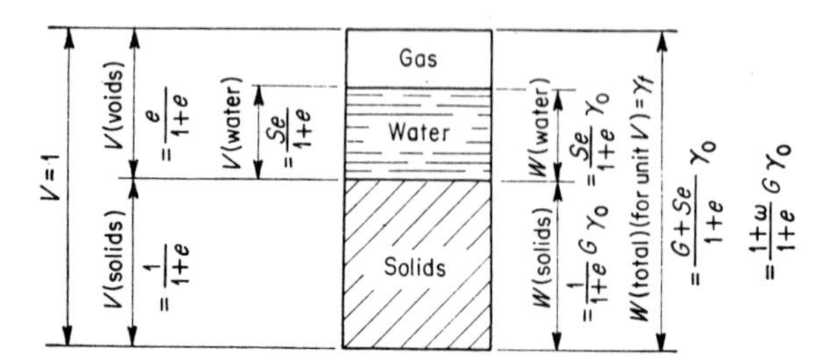

FIGURE 8.1 Relationship of weights and volumes in soil.

INTERNAL FRICTION AND COHESION

The angle of *internal friction* for a soil is expressed by

$$\tan \phi = \frac{\tau}{\sigma}$$

where ϕ = angle of internal friction

$\tan \phi$ = coefficient of internal friction

σ = normal force on given plane in cohesionless soil mass

τ = shearing force on same plane when sliding on plane is impending

For medium and coarse sands, the angle of internal friction is about 30° to 35°. The angle of internal friction for clays ranges from practically 0° to 20°.

The *cohesion* of a soil is the shearing strength that the soil possesses by virtue of its intrinsic pressure. The value of the ultimate cohesive resistance of a soil is usually designated by c. Average values for c are given in Table 8.2.

TABLE 8.2 Cohesive Resistance of Various Soil Types

	Cohesion c	
General soil type	lb/ft^2	(kPa)
Almost-liquid clay	100	(4.8)
Very soft clay	200	(9.6)
Soft clay	400	(19.1)
Medium clay	1000	(47.8)
Damp, muddy sand	400	(19.1)

VERTICAL PRESSURES IN SOILS

The vertical stress in a soil caused by a vertical, concentrated surface load may be determined with a fair degree of accuracy by the use of elastic theory. Two equations are in common use, the Boussinesq and the Westergaard. The Boussinesq equation applies to an elastic, isotropic, homogeneous mass that extends infinitely in all directions from a level surface. The vertical stress at a point in the mass is

$$\sigma_z = \frac{3P}{2\pi z^2}\left[1 + \left(\frac{r}{z}\right)^2\right]^{5/2}$$

The Westergaard equation applies to an elastic material laterally reinforced with horizontal sheets of negligible thickness and infinite rigidity, which prevent the mass from undergoing lateral strain. The vertical stress at a point in the mass, assuming a Poisson's ratio of zero, is

$$\sigma_z = \frac{P}{\pi z^2}\left[1 + 2\left(\frac{r}{z}\right)^2\right]^{3/2}$$

where σ_z = vertical stress at a point, lb/ft² (kPa)

P = total concentrated surface load, lb (N)

z = depth of point at which σ_z acts, measured vertically downward from surface, ft (m)

r = horizontal distance from projection of surface load P to point at which σ_z acts, ft (m)

For values of r/z between 0 and 1, the Westergaard equation gives stresses appreciably lower than those given by the Boussinesq equation. For values of r/z greater than 2.2, both equations give stresses less than $P/100z^2$.

LATERAL PRESSURES IN SOILS, FORCES ON RETAINING WALLS

The Rankine theory of lateral earth pressures, used for estimating approximate values for lateral pressures on retaining walls, assumes that the pressure on the back of a vertical wall is the same as the pressure that would exist on a vertical plane in an infinite soil mass. Friction between the wall and the soil is neglected. The pressure on a wall consists of (1) the lateral pressure of the soil held by the wall, (2) the pressure of the water (if any) behind the wall, and (3) the lateral pressure from any surcharge on the soil behind the wall.

Symbols used in this section are as follows:

γ = unit weight of soil, lb/ft^3 (kg/m^3) (saturated unit weight, dry unit weight, or buoyant unit weight, depending on conditions)

P = total thrust of soil, lb/linear ft (kg/linear m) of wall

H = total height of wall, ft (m)

ϕ = angle of internal friction of soil, degree

i = angle of inclination of ground surface behind wall with horizontal; also angle of inclination of line of action of total thrust P and pressures on wall with horizontal

K_A = coefficient of active pressure

K_P = coefficient of passive pressure
c = cohesion, lb/ft^2 (kPa)

LATERAL PRESSURE OF COHESIONLESS SOILS

For walls that retain cohesionless soils and are free to move an appreciable amount, the total thrust from the soil is

$$P = \frac{1}{2}\gamma H^2 \cos i \frac{\cos i - \sqrt{(\cos i)^2 - (\cos\phi)^2}}{\cos i + \sqrt{(\cos i)^2 - (\cos\phi)^2}}$$

When the surface behind the wall is level, the thrust is

$$P = \tfrac{1}{2}\gamma H^2 K_A$$

where
$$K_A = \left[\tan\left(45° - \frac{\phi}{2}\right)\right]^2$$

The thrust is applied at a point $H/3$ above the bottom of the wall, and the pressure distribution is triangular, with the maximum pressure of $2P/H$ occurring at the bottom of the wall.

For walls that retain cohesionless soils and are free to move only a slight amount, the total thrust is $1.12P$, where P is as given earlier. The thrust is applied at the midpoint of the wall, and the pressure distribution is trapezoidal, with the maximum pressure of $1.4P/H$ extending over the middle six-tenth of the height of the wall.

For walls that retain cohesionless soils and are completely restrained (very rare), the total thrust from the soil is

$$P = \frac{1}{2}\gamma H^2 \cos i \, \frac{\cos i + \sqrt{(\cos i)^2 - (\cos \phi)^2}}{\cos i - \sqrt{(\cos i)^2 - (\cos \phi)^2}}$$

When the surface behind the wall is level, the thrust is

$$P = {}^1\!/_2 \gamma H^2 K_P$$

where

$$K_P = \left[\tan\left(45° - \frac{\phi}{2}\right)\right]^2$$

The thrust is applied at a point $H/3$ above the bottom of the wall, and the pressure distribution is triangular, with the maximum pressure of $2P/H$ occurring at the bottom of the wall.

LATERAL PRESSURE OF COHESIVE SOILS

For walls that retain cohesive soils and are free to move a considerable amount over a long period of time, the total thrust from the soil (assuming a level surface) is

$$P = {}^1\!/_2 \gamma H^2 K_A - 2cH\sqrt{K_A}$$

or, because highly cohesive soils generally have small angles of internal friction,

$$P = {}^1\!/_2 \gamma H^2 - 2cH$$

The thrust is applied at a point somewhat below $H/3$ from the bottom of the wall, and the pressure distribution is approximately triangular.

For walls that retain cohesive soils and are free to move only a small amount or not at all, the total thrust from the soil is

$$P = \frac{1}{2}\gamma H^2 K_P$$

because the cohesion would be lost through plastic flow.

WATER PRESSURE

The total thrust from water retained behind a wall is

$$P = \frac{1}{2}\gamma_0 H^2$$

where H = height of water above bottom of wall, ft (m); and γ_0 = unit weight of water, lb/ft^3 (62.4 lb/ft^3 (1001 kg/m^3) for freshwater and 64 lb/ft^3 (1026.7 kg/m^3) for saltwater)

The thrust is applied at a point $H/3$ above the bottom of the wall, and the pressure distribution is triangular, with the maximum pressure of $2P/H$ occurring at the bottom of the wall. Regardless of the slope of the surface behind the wall, the thrust from water is always horizontal.

LATERAL PRESSURE FROM SURCHARGE

The effect of a surcharge on a wall retaining a cohesionless soil or an unsaturated cohesive soil can be accounted for by applying a uniform horizontal load of magnitude $K_A p$ over the entire height of the wall, where p is the surcharge in pound per square foot (kilopascal). For saturated cohesive soils, the full value of the surcharge p should be considered as acting over the entire height of the wall as a uniform horizontal load. K_A is defined earlier.

STABILITY OF SLOPES

Cohesionless Soils

A slope in a cohesionless soil without seepage of water is stable if

$$i < \phi$$

With seepage of water parallel to the slope, and assuming the soil to be saturated, an infinite slope in a cohesionless soil is stable if

$$\tan i < \left(\frac{\gamma_b}{\gamma_{sat}}\right) \tan \phi$$

where i = slope of ground surface

ϕ = angle of internal friction of soil

γ_b, γ_{sat} = unit weights, lb/ft^3 (kg/m^3)

Cohesive Soils

A slope in a cohesive soil is stable if

$$H < \frac{C}{\gamma N}$$

where H = height of slope, ft (m)

C = cohesion, lb/ft^2 (kg/m^2)

γ = unit weight, lb/ft^3 (kg/m^3)

N = stability number, dimensionless

For failure on the slope itself, without seepage water,

$$N = (\cos i)^2 (\tan i - \tan \phi)$$

Similarly, with seepage of water,

$$N = (\cos i)^2 \left[\tan i - \left(\frac{\gamma_b}{\gamma_{\text{sat}}} \right) \tan \phi \right]$$

When the slope is submerged, ϕ is the angle of internal friction of the soil and γ is equal to γ_b. When the surrounding water is removed from a submerged slope in a short time (sudden drawdown), ϕ is the weighted angle of internal friction, equal to $(\gamma_b/\gamma_{\text{sat}})\phi$, and γ is equal to γ_{sat}.

BEARING CAPACITY OF SOILS

The approximate ultimate bearing capacity under a long footing at the surface of a soil is given by Prandtl's equation as

$$q_u = \left(\frac{c}{\tan \phi} \right) + \frac{1}{2} \gamma_{\text{dry}} b \sqrt{K_p} \, (K_p e^{\pi \tan \phi} - 1)$$

where q_u = ultimate bearing capacity of soil, lb/ft^2 (kg/m^2)

c = cohesion, lb/ft^2 (kg/m^2)

ϕ = angle of internal friction, degree

γ_{dry} = unit weight of dry soil, lb/ft^3 (kg/m^3)

b = width of footing, ft (m)

d = depth of footing below surface, ft (m)

K_p = coefficient of passive pressure

$$= \left[\tan\left(45 + \frac{\phi}{2}\right)\right]^2$$

$e = 2.718 \cdots$

For footings below the surface, the ultimate bearing capacity of the soil may be modified by the factor $1 + Cd/b$. The coefficient C is about 2 for cohesionless soils and about 0.3 for cohesive soils. The increase in bearing capacity with depth for cohesive soils is often neglected.

SETTLEMENT UNDER FOUNDATIONS

The approximate relationship between loads on foundations and settlement is

$$\frac{q}{P} = C_1\left(1 + \frac{2d}{b}\right) + \frac{C_2}{b}$$

where q = load intensity, lb/ft^2 (kg/m^2)

P = settlement, in (mm)

d = depth of foundation below ground surface, ft (m)

b = width of foundation, ft (m)

C_1 = coefficient dependent on internal friction

C_2 = coefficient dependent on cohesion

The coefficients C_1 and C_2 are usually determined by bearing-plate loading tests.

SOIL COMPACTION TESTS

The *sand-cone method* is used to determine in the field the density of compacted soils in earth embankments, road fill, and structure backfill, as well as the density of natural soil deposits, aggregates, soil mixtures, or other similar materials. It is not suitable, however, for soils that are saturated, soft, or friable (crumble easily).

Characteristics of the soil are computed from

$$\text{Volume of soil, ft}^3\ (\text{m}^3) = \frac{\text{weight of sand filling hole, lb (kg)}}{\text{density of sand, lb/ft}^3\ (\text{kg/m}^3)}$$

$$\%\ \text{Moisture} = \frac{100(\text{weight of moist soil} - \text{weight of dry soil})}{\text{weight of dry soil}}$$

$$\text{Field density, lb/ft}^3\ (\text{kg/m}^3) = \frac{\text{weight of soil, lb (kg)}}{\text{volume of soil, ft}^3\ (\text{m}^3)}$$

$$\text{Dry density} = \frac{\text{field density}}{1 + \%\ \text{moisture}/100}$$

$$\%\ \text{Compaction} = \frac{100(\text{dry density})}{\text{max dry density}}$$

Maximum density is found by plotting a density–moisture curve.

Load-Bearing Test

One of the earliest methods for evaluating the *in situ* deformability of coarse-grained soils is the small-scale *load-bearing test*. Data developed from these tests have been used to provide a scaling factor to express the settlement ρ of a full-size footing from the settlement ρ_1 of a 1-ft^2 (0.0929-m^2) plate. This factor ρ/ρ_1 is given as a function of the width B of the full-size bearing plate as

$$\frac{\rho}{\rho^1} = \left(\frac{2B}{1 + B}\right)^2$$

From an elastic half-space solution, E_s' can be expressed from results of a plate load test in terms of the ratio of bearing pressure to plate settlement k_v as

$$E_s' = \frac{k_v(1 - \mu^2)\pi/4}{4B/(1 + B)^2}$$

where μ represents Poisson's ratio, usually considered to range between 0.30 and 0.40. The E_s' equation assumes that ρ_1 is derived from a rigid, 1-ft (0.3048-m)-diameter circular plate and that B is the equivalent diameter of the bearing area of a full-scale footing. Empirical formulations, such as the ρ/ρ_1 equation, may be significantly in error because of the limited footing-size range used and the large scatter of the database. Furthermore, consideration is not given to variations in the characteristics and stress history of the bearing soils.

California Bearing Ratio

The *California bearing ratio* (CBR) is often used as a measure of the quality of strength of a soil that underlies a pavement, for determining the thickness of the pavement, its base, and other layers.

$$\text{CBR} = \frac{F}{F_0}$$

where F = force per unit area required to penetrate a soil mass with a 3-in^2 (1935.6-mm^2) circular piston (about 2 in (50.8 mm) in diameter) at the rate of 0.05 in/min (1.27 mm/min); and F_0 = force per unit area required for corresponding penetration of a standard material.

Typically, the ratio is determined at 0.10-in (2.54-mm) penetration, although other penetrations sometimes are used. An excellent base course has a CBR of 100 percent. A compacted soil may have a CBR of 50 percent, whereas a weaker soil may have a CBR of 10.

Soil Permeability

The coefficient of permeability k is a measure of the rate of flow of water through saturated soil under a given hydraulic gradient *i*, cm/cm, and is defined in accordance with Darcy's law as

$$V = kiA$$

where V = rate of flow, cm^3/s, and A = cross-sectional area of soil conveying flow, cm^2.

Coefficient k is dependent on the grain-size distribution, void ratio, and soil fabric and typically may vary from as much as 10 cm/s for gravel to less than 10^{-7} for clays. For typical soil deposits, k for horizontal flow is greater than k for vertical flow, often by an order of magnitude.

COMPACTION EQUIPMENT

A wide variety of equipment is used to obtain compaction in the field. Sheepsfoot rollers generally are used on soils that contain high percentages of clay. Vibrating rollers are used on more granular soils.

To determine maximum depth of lift, make a test fill. In the process, the most suitable equipment and pressure to be applied, lb/in^2 (kPa), for ground contact also can be determined. Equipment selected should be able to produce desired compaction with four to eight passes. Desirable speed of rolling also can be determined. Average speeds, mi/h (km/h), under normal conditions are given in Table 8.3.

Compaction production can be computed from

$$\text{yd}^3/\text{h}\ (\text{m}^3/\text{h}) = \frac{16WSLFE}{P}$$

where W = width of roller, ft (m)

S = roller speed, mi/h (km/h)

TABLE 8.3 Average Speeds of Rollers

Type	mi/h	(km/h)
Grid rollers	12	(19.3)
Sheepsfoot rollers	3	(4.8)
Tamping rollers	10	(16.1)
Pneumatic rollers	8	(12.8)

L = lift thickness, in (mm)

F = ratio of pay yd^3 (m^3) to loose yd^3 (m^3)

E = efficiency factor (allows for time losses, such as those due to turns): 0.90, excellent; 0.80, average; 0.75, poor

P = number of passes

FORMULAS FOR EARTHMOVING

External forces offer *rolling resistance* to the motion of wheeled vehicles, such as tractors and scrapers. The engine has to supply power to overcome this resistance; the greater the resistance is, the more power needed to move a load. Rolling resistance depends on the weight on the wheels and the tire penetration into the ground:

$$R = R_f W + R_p p W \tag{8.6}$$

where R = rolling resistance, lb (N)

R_f = rolling-resistance factor, lb/ton (N/tonne)

W = weight on wheels, ton (tonne)

R_p = tire-penetration factor, lb/ton·in (N/tonne·mm) penetration

p = tire penetration, in (mm)

R_f usually is taken as 40 lb/ton (or 2 percent lb/lb) (173 N/t) and R_p as 30 lb/ton·in (1.5% lb/lb·in) (3288 N/t·mm). Hence, Eq. (8.6) can be written as

$$R = (2\% + 1.5\%p)\, W' = R'W' \tag{8.7}$$

where W' = weight on wheels, lb (N); and $R' = 2\% + 1.5\%p$.

Additional power is required to overcome rolling resistance on a slope. Grade resistance also is proportional to weight:

$$G = R_g sW \tag{8.8}$$

where G = grade resistance, lb (N)

R_g = grade-resistance factor = 20 lb/ton (86.3 N/t) = 1% lb/lb (N/N)

s = percent grade, positive for uphill motion, negative for downhill

Thus, the total road resistance is the algebraic sum of the rolling and grade resistances, or the total pull, lb (N), required:

$$T = (R' + R_g s)W' = (2\% + 1.5\%p + 1\%s)W' \tag{8.9}$$

In addition, an allowance may have to be made for loss of power with altitude. If so, allow 3 percent pull loss for each 1000 ft (305 m) above 2500 ft (762 m).

Usable pull P depends on the weight W on the drivers:

$$P = fW \tag{8.10}$$

where f = coefficient of traction.

Earth Quantities Hauled

When soils are excavated, they increase in volume, or swell, because of an increase in voids:

$$V_b = V_L L = \frac{100}{100 + \%\ \text{swell}} V_L \tag{8.11}$$

where V_b = original volume, yd^3 (m^3), or bank yards

V_L = loaded volume, yd^3 (m^3), or loose yards

L = load factor

When soils are compacted, they decrease in volume:

$$V_c = V_b S \tag{8.12}$$

where V_c = compacted volume, yd^3 (m^3); and S = shrinkage factor.

Bank yards moved by a hauling unit equals weight of load, lb (kg), divided by density of the material in place, lb (kg), per bank yard (m^3).

SCRAPER PRODUCTION

Production is measured in terms of tons or bank cubic yards (cubic meters) of material a machine excavates and discharges, under given job conditions, in 1 h.

$$\text{Production, bank yd}^3\text{/h (m}^3\text{/h)} = \text{load, yd}^3 \text{ (m}^3\text{)} \times \text{trips per hour}$$

$$\text{Trips per hour} = \frac{\text{working time, min/h}}{\text{cycle time, min}}$$

The load, or amount of material a machine carries, can be determined by weighing or estimating the volume. Payload estimating involves determination of the bank cubic yards (cubic meters) being carried, whereas the excavated material expands when loaded into the machine. For determination of bank cubic yards (cubic meters) from loose volume, the amount of swell or the load factor must be known.

Weighing is the most accurate method of determining the actual load. This is normally done by weighing one wheel or axle at a time with portable scales, adding the wheel or axle weights, and subtracting the weight empty. To reduce error, the machine should be relatively level. Enough loads should be weighed to provide a good average:

$$\text{Bank yd}^3 = \frac{\text{weight of load, lb (kg)}}{\text{density of material, lb/bank yd}^3\text{ (kg/m}^3\text{)}}$$

Equipment Required

To determine the number of scrapers needed on a job, required production must first be computed:

$$\text{Production required, yd}^3\text{/h (m}^3\text{/h)} = \frac{\text{quantity, bank yd}^3\text{ (m}^3\text{)}}{\text{working time, h}}$$

No. of scrapers needed

$$= \frac{\text{production required, yd}^3\text{/h (m}^3\text{/h)}}{\text{production per unit, yd}^3\text{/h (m}^3\text{/h)}}$$

No. of scrapers a pusher can load

$$= \frac{\text{scraper cycle time, min}}{\text{pusher cycle time, min}}$$

Because speeds and distances may vary on haul and return, haul and return times are estimated separately.

Variable time, min

$$= \frac{\text{haul distance, ft}}{88 \times \text{speed, mi/h}} + \frac{\text{return distance, ft}}{88 \times \text{speed, mi/h}}$$

$$= \frac{\text{haul distance, m}}{16.7 \times \text{speed, km/h}} + \frac{\text{return distance, m}}{16.7 \times \text{speed, km/h}}$$

Haul speed may be obtained from the equipment specification sheet when the drawbar pull required is known.

VIBRATION CONTROL IN BLASTING

Explosive users should take steps to minimize vibration and noise from blasting and protect themselves against damage claims.

Vibrations caused by blasting are propagated with a velocity V, ft/s (m/s), frequency f, Hz, and wavelength L, ft (m), related by

$$L = \frac{V}{f}$$

Velocity v, in/s (mm/s), of the particles disturbed by the vibrations depends on the amplitude of the vibrations A, in (mm):

$$v = 2\pi f A$$

If the velocity v_1 at a distance D_1 from the explosion is known, the velocity v_2 at a distance D_2 from the explosion may be estimated from

$$v_2 \approx v_1 \left(\frac{D_1}{D_2}\right)^{1.5}$$

The acceleration a, in/s^2 (mm/s^2), of the particles is given by

$$a = 4\pi^2 f^2 A$$

For a charge exploded on the ground surface, the overpressure P, lb/in^2 (kPa), may be computed from

$$P = 226.62 \left(\frac{W^{1/3}}{D}\right)^{1.407}$$

where W = maximum weight of explosives, lb (kg) per delay; and D = distance, ft (m), from explosion to exposure.

The sound pressure level, decibels, may be computed from

$$dB = \left(\frac{P}{6.95 \times 10^{-28}}\right)^{0.084}$$

For vibration control, blasting should be controlled with the scaled-distance formula:

$$v = H\left(\frac{D}{\sqrt{W}}\right)^{-\beta}$$

where β = constant (varies for each site), and H = constant (varies for each site).

Distance to exposure, ft (m), divided by the square root of maximum pounds (kg) per delay is known as *scaled distance.*

Most courts have accepted the fact that a particle velocity not exceeding 2 in/s (50.8 mm/s) does not damage any part of any structure. This implies that, for this velocity, vibration damage is unlikely at scaled distances larger than 8.

CHAPTER 9

BUILDING AND STRUCTURES FORMULAS

LOAD-AND-RESISTANCE FACTOR DESIGN FOR SHEAR IN BUILDINGS

Based on the *American Institute of Steel Construction* (AISC) specifications for *load-and-resistance factor design* (LRFD) for buildings, the shear capacity V_u, kip (kN = 4.448 × kip), of flexural members may be computed from the following:

$$V_u = 0.54 F_{yw} A_w \qquad \text{when} \qquad \frac{h}{t_w} \leq \alpha$$

$$V_u = \frac{0.54 \alpha F_{yw} A_w}{h/t_w} \qquad \text{when} \qquad \alpha < \frac{h}{t_w} \leq 1.25\alpha$$

$$V_u = \frac{23{,}760 k A_w}{(h/t_w)^2} \qquad \text{when} \qquad \frac{h}{t_w} > 1.25\alpha$$

where F_{yw} = specified minimum yield stress of web, ksi (MPa = 6.894 × ksi)

A_w = web area, in² (mm²) = dt_w

$\alpha = 187\sqrt{k/F_{yw}}$

k = 5 if a/h exceeds 3.0 or $67{,}600/(h/t_w)^2$, or if stiffeners are not required

= $5 + 5/(a/h)^2$, otherwise

Stiffeners are required when the shear exceeds V_u. In unstiffened girders, h/t_w may not exceed 260. In girders with stiffeners, maximum h/t_w permitted is $2{,}000/\sqrt{F_{yf}}$ for $a/h \leq 1.5$ or $14{,}000/\sqrt{F_{yf}(F_{yf} + 16.5)}$ for $a/h > 1.5$, where F_{yf} is the specified minimum yield stress, ksi, of the flange.

For shear capacity with tension-field action, see the AISC specification for LRFD.

ALLOWABLE-STRESS DESIGN FOR BUILDING COLUMNS

The AISC specification for *allowable-stress design* (ASD) for buildings provides two formulas for computing allowable compressive stress F_a, ksi (MPa), for main members. The formula to use depends on the relationship of the largest effective slenderness ratio Kl/r of the cross section of any unbraced segment to a factor C_c defined by the following equation and Table 9.1:

$$C_c = \sqrt{\frac{2\pi^2 E}{F_y}} = \frac{756.6}{\sqrt{F_y}}$$

where E = modulus of elasticity of steel

= 29,000 ksi (128.99 GPa)

F_y = yield stress of steel, ksi (MPa)

When Kl/r is less than C_c,

TABLE 9.1 Values of C_c

F_y	C_c
36	126.1
50	107.0

$$F_a = \frac{\left[1 - \frac{(Kl/r)^2}{2C_c^2}\right] F_y}{\text{F.S.}}$$

where F.S. = safety factor $= \frac{5}{3} + \frac{3(Kl/r)}{8C_c} - \frac{(Kl/r)^3}{8C_c^3}$.

When Kl/r exceeds C_c,

$$F_a = \frac{12\pi^2 E}{23(Kl/r)^2} = \frac{150{,}000}{(Kl/r)^2}$$

The effective-length factor K, equal to the ratio of effective-column length to actual unbraced length, may be greater or less than 1.0. Theoretical K values for six idealized conditions, in which joint rotation and translation are either fully realized or nonexistent, are tabulated in Fig. 9.1.

	(a)	(b)	(c)	(d)	(e)	(f)
BUCKLED SHAPE OF COLUMN IS SHOWN BY DASHED LINE	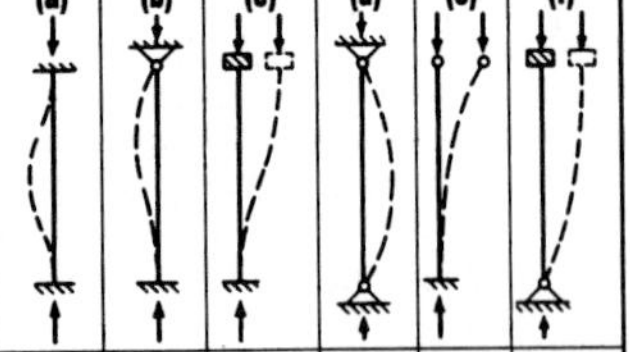					
THEORETICAL K VALUE	0.5	0.7	1.0	1.0	2.0	2.0
RECOMMENDED DESIGN VALUE WHEN IDEAL CONDITIONS ARE APPROXIMATED	0.65	0.80	1.2	1.0	2.10	2.0
END CONDITION		ROTATION FIXED AND TRANSLATION FIXED ROTATION FREE AND TRANSLATION FIXED ROTATION FIXED AND TRANSLATION FREE ROTATION FREE AND TRANSLATION FREE				

FIGURE 9.1 Values of effective-length factor K for columns.

LOAD-AND-RESISTANCE FACTOR DESIGN FOR BUILDING COLUMNS

Plastic analysis of prismatic compression members in buildings is permitted if $\sqrt{F_y}\,(l/r)$ does not exceed 800 and $F_u \leq 65$ ksi (448 MPa). For axially loaded members with $b/t \leq \lambda_r$, the maximum load P_u, ksi (MPa = 6.894 × ksi), may be computed from

$$P_u = 0.85 A_g F_{cr}$$

where A_g = gross cross-sectional area of the member

$$F_{cr} = 0.658^{\lambda} F_y \text{ for } \lambda \leq 2.25$$

$$= 0.877\, F_y/\lambda \text{ for } \lambda > 2.25$$

$$\lambda = (Kl/r)^2 (F_y/286{,}220)$$

The AISC specification for LRFD presents formulas for designing members with slender elements.

ALLOWABLE-STRESS DESIGN FOR BUILDING BEAMS

The maximum fiber stress in bending for laterally supported beams and girders is $F_b = 0.66F_y$ if they are compact, except for hybrid girders and members with yield points exceeding 65 ksi (448.1 MPa). $F_b = 0.60F_y$ for noncompact sections. F_y is the minimum specified yield strength of the steel, ksi (MPa). Table 9.2 lists values of F_b for two grades of steel.

TABLE 9.2 Allowable Bending Stresses in Braced Beams for Buildings

Yield strength, ksi (MPa)	Compact, $0.66F_y$ (MPa)	Noncompact, $0.60F_y$ (MPa)
36 (248.2)	24 (165.5)	22 (151.7)
50 (344.7)	33 (227.5)	30 (206.8)

The allowable extreme-fiber stress of $0.60F_y$ applies to laterally supported, unsymmetrical members, except channels, and to noncompact box sections. Compression on outer surfaces of channels bent about their major axis should not exceed $0.60F_y$ or the value given by Eq. (9.5).

The allowable stress of $0.66F_y$ for compact members should be reduced to $0.60F_y$ when the compression flange is unsupported for a length, in (mm), exceeding the smaller of

$$l_{\max} = \frac{76.0b_f}{\sqrt{F_y}} \tag{9.1}$$

$$l_{\max} = \frac{20{,}000}{F_y d/A_f} \tag{9.2}$$

where b_f = width of compression flange, in (mm)

d = beam depth, in (mm)

A_f = area of compression flange, in^2 (mm^2)

The allowable stress should be reduced even more when l/r_T exceeds certain limits, where l is the unbraced length, in (mm), of the compression flange, and r_T is the radius of gyration, in (mm), of a portion of the beam consisting of the compression flange and one-third of the part of the web in compression.

For $\sqrt{102{,}000C_b/F_y} \leq l/r_T \leq \sqrt{510{,}00C_b/F_y}$, use

$$F_b = \left[\frac{2}{3} - \frac{F_y(l/r_T)^2}{1{,}530{,}000C_b}\right]F_y \tag{9.3}$$

For $l/r_T > \sqrt{510{,}000C_b/F_y}$, use

$$F_b = \frac{170{,}000C_b}{(l/r_T)^2} \tag{9.4}$$

where C_b = modifier for moment gradient (Eq. 9.6).

When, however, the compression flange is solid and nearly rectangular in cross section, and its area is not less than that of the tension flange, the allowable stress may be taken as

$$F_b = \frac{12{,}000C_b}{ld/A_f} \tag{9.5}$$

When Eq. (9.5) applies (except for channels), F_b should be taken as the larger of the values computed from Eqs. (9.5) and (9.3) or (9.4), but not more than $0.60F_y$.

The moment-gradient factor C_b in Eqs. (9.1) to (9.5) may be computed from

$$C_b = 1.75 + 1.05\frac{M_1}{M_2} + 0.3\left(\frac{M_1}{M_2}\right)^2 \leq 2.3 \tag{9.6}$$

where M_1 = smaller beam end moment, and M_2 = larger beam end moment.

The algebraic sign of M_1/M_2 is positive for double-curvature bending and negative for single-curvature bending. When the bending moment at any point within an unbraced length is larger than that at both ends, the value of C_b should be taken as unity. For braced frames, C_b should be taken as unity for computation of F_{bx} and F_{by}.

Equations (9.4) and (9.5) can be simplified by introducing a new term:

$$Q = \frac{(l/r_T)^2 F_y}{510{,}000 C_b} \tag{9.7}$$

Now, for $0.2 \le Q \le 1$,

$$F_b = \frac{(2 - Q)F_y}{3} \tag{9.8}$$

For $Q > 1$:

$$F_b = \frac{F_y}{3Q} \tag{9.9}$$

As for the preceding equations, when Eq. (9.1) applies (except for channels), F_b should be taken as the largest of the values given by Eqs. (9.1) and (9.8) or (9.9), but not more than $0.60F_y$.

LOAD-AND-RESISTANCE FACTOR DESIGN FOR BUILDING BEAMS

For a compact section bent about the major axis, the unbraced length L_b of the compression flange, where plastic hinges may form at failure, may not exceed L_{pd}, given by Eqs. (9.10) and (9.11) that follow. For beams bent about the minor axis and square and circular beams, L_b is not restricted for plastic analysis.

For I-shaped beams, symmetrical about both the major and the minor axis or symmetrical about the minor axis but with the compression flange larger than the tension flange, including hybrid girders, loaded in the plane of the web:

$$L_{pd} = \frac{3600 + 2200(M_1/M_p)}{F_{yc}} r_y \tag{9.10}$$

where F_{yc} = minimum yield stress of compression flange, ksi (MPa)

M_1 = smaller of the moments, in·kip (mm·MPa) at the ends of the unbraced length of beam

M_p = plastic moment, in·kip (mm·MPa)

r_y = radius of gyration, in (mm), about minor axis

The plastic moment M_p equals $F_y Z$ for homogeneous sections, where Z = plastic modulus, in^3 (mm^3); and for hybrid girders, it may be computed from the fully plastic distribution. M_1/M_p is positive for beams with reverse curvature.

For solid rectangular bars and symmetrical box beams:

$$L_{pd} = \frac{5000 + 3000(M_1/M_p)}{F_y} r_y \geq 3000 \frac{r_y}{F_y} \tag{9.11}$$

The *flexural design strength* $0.90M_n$ is determined by the limit state of lateral-torsional buckling and should be calculated for the region of the last hinge to form and for regions not adjacent to a plastic hinge. The specification gives formulas for M_n that depend on the geometry of the section and the bracing provided for the compression flange.

For compact sections bent about the major axis, for example, M_n depends on the following unbraced lengths:

L_b = the distance, in (mm), between points braced against lateral displacement of the compression flange or between points braced to prevent twist

L_p = limiting laterally unbraced length, in (mm), for full plastic-bending capacity

= $300r_y/\sqrt{F_{yf}}$ for I shapes and channels

= $3750(r_y/M_p)/\sqrt{JA}$ for solid rectangular bars and box beams

F_{yf} = flange yield stress, ksi (MPa)

J = torsional constant, in^4 (mm^4) (see AISC "Manual of Steel Construction" on LRFD)

A = cross-sectional area, in^2 (mm^2)

L_r = limiting laterally unbraced length, in (mm), for inelastic lateral buckling

For I-shaped beams symmetrical about the major or the minor axis, or symmetrical about the minor axis with the compression flange larger than the tension flange and channels loaded in the plane of the web:

$$L_r = \frac{r_y x_1}{F_{yw} - F_r}\sqrt{1 + \sqrt{1 + X_2 F_L^2}} \qquad (9.12)$$

where F_{yw} = specified minimum yield stress of web, ksi (MPa)

F_r = compressive residual stress in flange

= 10 ksi (68.9 MPa) for rolled shapes, 16.5 ksi (113.6 MPa), for welded sections

F_L = smaller of $F_{yf} - F_r$ or F_{yw}

F_{yf} = specified minimum yield stress of flange, ksi (MPa)

$X_1 = (\pi/S_x)\sqrt{EGJA/2}$

$X_2 = (4C_w/I_y)\,(S_x/GJ)^2$

E = elastic modulus of the steel

G = shear modulus of elasticity

S_x = section modulus about major axis, in^3 (mm^3) (with respect to the compression flange if that flange is larger than the tension flange)

C_w = warping constant, in^6 (mm^6) (see AISC manual on LRFD)

I_y = moment of inertia about minor axis, in^4 (mm^4)

For the previously mentioned shapes, the limiting buckling moment M_r, ksi (MPa), may be computed from

$$M_r = F_L S_x \tag{9.13}$$

For compact beams with $L_b \leq L_r$, bent about the major axis:

$$M_n = C_b\left[M_p - (M_p - M_r)\frac{L_b - L_p}{L_r - L_p}\right] \leq M_p \tag{9.14}$$

where $C_b = 1.75 + 1.05(M_1/M_2) + 0.3(M_1/M_2) \leq 2.3$, where M_1 is the smaller and M_2 the larger end moment in the unbraced segment of the beam; M_1/M_2 is positive for reverse curvature and equals 1.0 for unbraced cantilevers and beams with moments over much of the unbraced segment equal to or greater than the larger of the segment end moments.

(See Galambos, T. V., *Guide to Stability Design Criteria for Metal Structures*, 4th ed., John Wiley & Sons, New York, for use of larger values of C_b.)

For solid rectangular bars bent about the major axis:

$$L_r = 57{,}000\left(\frac{r_y}{M_r}\right)\sqrt{JA} \tag{9.15}$$

and the limiting buckling moment is given by:

$$M_r = F_y S_x \tag{9.16}$$

For symmetrical box sections loaded in the plane of symmetry and bent about the major axis, M_r should be determined from Eq. (9.13) and L_r from Eq. (9.15)

For compact beams with $L_b > L_r$, bent about the major axis:

$$M_n = M_{cr} \leq C_b M_r \tag{9.17}$$

where M_{cr} = critical elastic moment, kip·in (MPa·mm).

For shapes to which Eq. (9.17) applies:

$$M_{cr} = C_b \frac{\pi}{L_b}\sqrt{EI_y GJ + I_y C_w \left(\frac{\pi E}{L_b}\right)^2} \tag{9.18}$$

For solid rectangular bars and symmetrical box sections:

$$M_{cr} = \frac{57{,}000 C_b \sqrt{JA}}{L_b / r_y} \tag{9.19}$$

For determination of the flexural strength of noncompact plate girders and other shapes not covered by the preceding requirements, see the AISC manual on LRFD.

ALLOWABLE-STRESS DESIGN FOR SHEAR IN BUILDINGS

The AISC specification for ASD specifies the following allowable shear stresses F_v, ksi (ksi × 6.894 = MPa):

$$F_v = 0.40F_y \qquad \frac{h}{t_w} \leq \frac{380}{\sqrt{F_y}}$$

$$F_v = \frac{C_v F_y}{289} \leq 0.40F_y \qquad \frac{h}{t_w} > \frac{380}{\sqrt{F_y}}$$

where

$$C_v = 45{,}000k_v/F_y(h/t_w)^2 \qquad \text{for } C_v < 0.8$$

$$= \sqrt{36{,}000k_v/F_y(h/t_w)^2} \qquad \text{for } C_v > 0.8$$

$$k_v = 4.00 + 5.34/(a/h)^2 \qquad \text{for } a/h < 1.0$$

$$= 5.34 + 4.00/(a/h)^2 \qquad \text{for } a/h > 1.0$$

a = clear distance between transverse stiffeners

The allowable shear stress with tension-field action is

$$F_v = \frac{F_y}{289}\left[C_v + \frac{1 - C_v}{1.15\sqrt{1 + (a/h)^2}}\right] \leq 0.40F_y$$

where $C_v \leq 1$

When the shear in the web exceeds F_v, stiffeners are required.

Within the boundaries of a rigid connection of two or more members with webs lying in a common plane, shear stresses in the webs generally are high. The commentary on the AISC specification for buildings states that such webs should be reinforced when the calculated shear stresses,

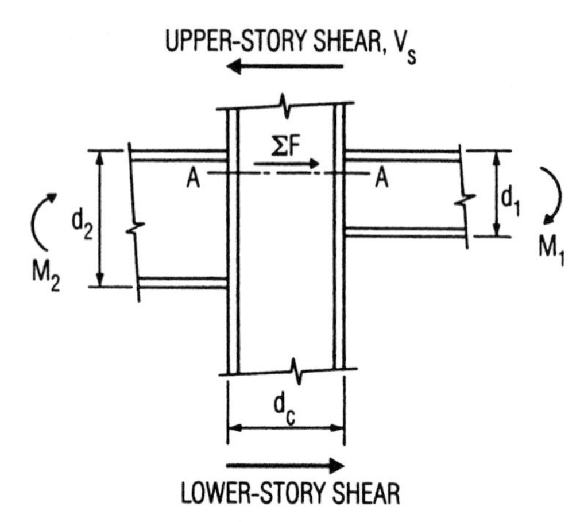

FIGURE 9.2 Rigid connection of steel members with webs in a common plane.

such as those along plane *AA* in Fig. 9.2, exceed F_v, that is, when ΣF is larger than $d_c t_w F_v$, where d_c is the depth and t_w is the web thickness of the member resisting ΣF. The shear may be calculated from

$$\Sigma F = \frac{M_1}{0.95d_1} + \frac{M_2}{0.95d_2} - V_s$$

where V_s = shear on the section

$M_1 = M_{1L} + M_{1G}$

M_{1L} = moment due to the gravity load on the leeward side of the connection

M_{1G} = moment due to the lateral load on the leeward side of the connection

$$M_2 = M_{2L} - M_{2G}$$

M_{2L} = moment due to the lateral load on the windward side of the connection

M_{2G} = moment due to the gravity load on the windward side of the connection

STRESSES IN THIN SHELLS

Results of membrane and bending theories are expressed in terms of unit forces and unit moments, acting per unit of length over the thickness of the shell. To compute the unit stresses from these forces and moments, usual practice is to assume normal forces and shears to be uniformly distributed over the shell thickness and bending stresses to be linearly distributed.

Then, normal stresses can be computed from equations of the form:

$$f_x = \frac{N_x}{t} + \frac{M_x}{t^3/12} z \qquad (9.20)$$

where z = distance from middle surface

t = shell thickness

M_x = unit bending moment about an axis parallel to direction of unit normal force N_x

Similarly, shearing stresses produced by central shears T and twisting moments D may be calculated from equations of the form:

$$v_{xy} = \frac{T}{t} \pm \frac{D}{t^3/12} z \tag{9.21}$$

Normal shearing stresses may be computed on the assumption of a parabolic stress distribution over the shell thickness:

$$v_{xz} = \frac{V}{t^3/6} \left(\frac{t^2}{4} - z^2 \right) \tag{9.22}$$

where V = unit shear force normal to middle surface.

BEARING PLATES

To resist a beam reaction, the minimum bearing length N in the direction of the beam span for a bearing plate is determined by equations for prevention of local web yielding and web crippling. A larger N is generally desirable but is limited by the available wall thickness.

When the plate covers the full area of a concrete support, the area, in^2 (mm^2), required by the bearing plate is

$$A_1 = \frac{R}{0.35 f_c'}$$

where R = beam reaction, kip (kN), f_c' = specified compressive strength of the concrete, ksi (MPa). When the plate covers less than the full area of the concrete support, then, as determined from Table 9.3,

TABLE 9.3 Allowable Bearing Stress, F_p, on Concrete and Masonry[†]

Full area of concrete support	$0.35f'_c$
Less than full area of concrete support	$0.35f'_c\sqrt{\dfrac{A_1}{A_2}} \leq 0.70f'_c$
Sandstone and limestone	0.40
Brick in cement mortar	0.25

[†] Units in MPa = 6.895 × ksi.

$$A_1 = \left(\frac{R}{0.35\, f'_c\, \sqrt{A_2}}\right)^2$$

where A_2 = full cross-sectional area of concrete support, in^2 (mm^2).

With N established, usually rounded to full inches (millimeters), the minimum width of plate B, in (mm), may be calculated by dividing A_1 by N and then rounded off to full inches (millimeters), so that $BN \geq A_1$. Actual bearing pressure f_p, ksi (MPa), under the plate then is

$$f_p = \frac{R}{BN}$$

The plate thickness usually is determined with the assumption of cantilever bending of the plate:

$$t = \left(\frac{1}{2}B - k\right)\sqrt{\frac{3f_p}{F_b}}$$

where t = minimum plate thickness, in (mm)

k = distance, in (mm), from beam bottom to top of web fillet

F_b = allowable bending stress of plate, ksi (MPa)

COLUMN BASE PLATES

The area A_1, in^2 (mm^2), required for a base plate under a column supported by concrete should be taken as the larger of the values calculated from the equation cited earlier, with R taken as the total column load, kip (kN), or

$$A_1 = \frac{R}{0.70 f'_c}$$

Unless the projections of the plate beyond the column are small, the plate may be designed as a cantilever assumed to be fixed at the edges of a rectangle with sides equal to $0.80b$ and $0.95d$, where b is the column flange width, in (mm), and d is the column depth, in (mm).

To minimize material requirements, the plate projections should be nearly equal. For this purpose, the plate length N, in (mm) (in the direction of d), may be taken as

$$N = \sqrt{A_1} + 0.5(0.95d - 0.80b)$$

The width B, in (mm), of the plate then may be calculated by dividing A_1 by N. Both B and N may be selected in full inches (millimeters) so that $BN \geq A_1$. In that case, the bearing pressure f_p, ksi (MPa), may be determined from the preceding equation. Thickness of plate, determined by cantilever bending, is given by

$$t = 2p\sqrt{\frac{f_p}{F_y}}$$

where F_y = minimum specified yield strength, ksi (MPa), of plate; and p = larger of $0.5(N - 0.95d)$ and $0.5(B - 0.80b)$.

When the plate projections are small, the area A_2 should be taken as the maximum area of the portion of the supporting surface that is geometrically similar to and concentric with the loaded area. Thus, for an H-shaped column, the column load may be assumed distributed to the concrete over an H-shaped area with flange thickness L, in (mm), and web thickness $2L$:

$$L = \frac{1}{4}(d + b) - \frac{1}{4}\sqrt{(d + b)^2 - \frac{4R}{F_p}}$$

where F_p = allowable bearing pressure, ksi (MPa), on support. (If L is an imaginary number, the loaded portion of the supporting surface may be assumed rectangular as discussed earlier.) Thickness of the base plate should be taken as the larger of the values calculated from the preceding equation and

$$t = L\sqrt{\frac{3f_p}{F_b}}$$

BEARING ON MILLED SURFACES

In building construction, allowable bearing stress for milled surfaces, including bearing stiffeners, and pins in reamed, drilled, or bored holes, is $F_p = 0.90F_y$, where F_y is the yield strength of the steel, ksi (MPa).

For expansion rollers and rockers, the allowable bearing stress, kip/linear in (kN/mm), is

$$F_p = \frac{F_y - 13}{20} 0.66d$$

where d is the diameter, in (mm), of the roller or rocker. When parts in contact have different yield strengths, F_y is the smaller value.

PLATE GIRDERS IN BUILDINGS

For greatest resistance to bending, as much of a plate girder cross section as practicable should be concentrated in the flanges, at the greatest distance from the neutral axis. This might require, however, a web so thin that the girder would fail by web buckling before it reached its bending capacity. To preclude this, the AISC specification limits h/t.

For an unstiffened web, this ratio should not exceed

$$\frac{h}{t} = \frac{14{,}000}{\sqrt{F_y(F_y + 16.5)}}$$

where F_y = yield strength of compression flange, ksi (MPa).

Larger values of h/t may be used, however, if the web is stiffened at appropriate intervals.

For this purpose, vertical angles may be fastened to the web or vertical plates welded to it. These transverse stiffeners are not required, though, when h/t is less than the value computed from the preceding equation or Table 9.4.

TABLE 9.4 Critical h/t for Plate Girders in Buildings

F_y, ksi	(MPa)	$\frac{14{,}000}{\sqrt{F_y(F_y + 16.5)}}$	$\frac{2{,}000}{\sqrt{F_y}}$
36	(248)	322	333
50	(345)	243	283

With transverse stiffeners spaced not more than 1.5 times the girder depth apart, the web clear-depth/thickness ratio may be as large as

$$\frac{h}{t} = \frac{2000}{\sqrt{F_y}}$$

If, however, the web depth/thickness ratio h/t exceeds $760/\sqrt{F_b}$, where F_b, ksi (MPa), is the allowable bending stress in the compression flange that would ordinarily apply, this stress should be reduced to F'_b, given by the following equations:

$$F'_b = R_{PG} R_e F_b$$

$$R_{PG} = \left[1 - 0.0005\frac{A_w}{A_f}\left(\frac{h}{t} - \frac{760}{\sqrt{F_b}}\right)\right] \leq 1.0$$

$$R_e = \left[\frac{12 + (A_w/A_f)(3\alpha - \alpha^3)}{12 + 2(A_w/A_f)}\right] \leq 1.0$$

where A_w = web area, in^2 (mm^2)

A_f = area of compression flange, in^2 (mm^2)

$$\alpha = 0.6F_{yw}/F_b \leq 1.0$$

F_{yw} = minimum specified yield stress, ksi, (MPa), of web steel

In a hybrid girder, where the flange steel has a higher yield strength than the web, the preceding equation protects against excessive yielding of the lower strength web in the vicinity of the higher strength flanges. For nonhybrid girders, $R_e = 1.0$.

LOAD DISTRIBUTION TO BENTS AND SHEAR WALLS

Provision should be made for all structures to transmit lateral loads, such as those from wind, earthquakes, and traction and braking of vehicles, to foundations and their supports that have high resistance to displacement. For this purpose, various types of bracing may be used, including struts, tension ties, diaphragms, trusses, and shear walls.

Deflections of Bents and Shear Walls

Horizontal deflections in the planes of bents and shear walls can be computed on the assumption that they act as cantilevers. Deflections of braced bents can be calculated by the dummy-unit-load method or a matrix method. Deflections of rigid frames can be computed by adding the drifts of the stories, as determined by moment distribution or a matrix method.

For a shear wall (Fig. 9.3), the deflection in its plane induced by a load in its plane is the sum of the flexural

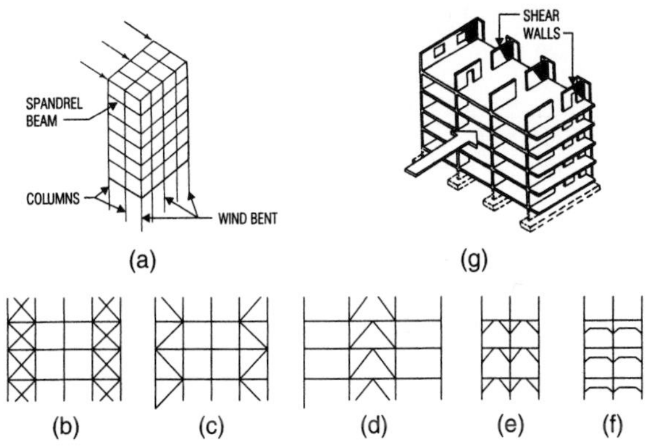

FIGURE 9.3 Building frame resists lateral forces with (*a*) wind bents or (*g*) shear walls or a combination of the two. Bents may be braced in any of several ways, including (*b*) X bracing, (*c*) K bracing, (*d*) inverted V bracing, (*e*) knee bracing, and (*f*) rigid connections.

deflection as a cantilever and the deflection due to shear. Thus, for a wall with solid rectangular cross section, the deflection at the top due to uniform load is

$$\delta = \frac{1.5wH}{Et}\left[\left(\frac{H}{L}\right)^3 + \frac{H}{L}\right]$$

where w = uniform lateral load

H = height of the wall

E = modulus of elasticity of the wall material

t = wall thickness

L = length of wall

For a shear wall with a concentrated load P at the top, the deflection at the top is

$$\delta_c = \frac{4P}{Et}\left[\left(\frac{H}{L}\right)^3 + 0.75\frac{H}{L}\right]$$

If the wall is fixed against rotation at the top, however, the deflection is

$$\delta_f = \frac{P}{Et}\left[\left(\frac{H}{L}\right)^3 + 3\frac{H}{L}\right]$$

Units used in these equations are those commonly applied in *United States Customary System* (USCS) and the *System International* (SI) measurements, that is, kip (kN), lb/in^2 (MPa), ft (m), and in (mm).

Where shear walls contain openings, such as those for doors, corridors, or windows, computations for deflection and rigidity are more complicated. Approximate methods, however, may be used.

COMBINED AXIAL COMPRESSION OR TENSION AND BENDING

The AISC specification for allowable stress design for buildings includes three interaction formulas for combined axial compression and bending.

When the ratio of computed axial stress to allowable axial stress f_u/F_a exceeds 0.15, both of the following equations must be satisfied:

$$\frac{f_a}{F_a} + \frac{C_{mx} f_{bx}}{(1 - f_a / F'_{ex}) F_{bx}} + \frac{C_{my} f_{by}}{(1 - f_a / F'_{ey}) F_{by}} \leq 1$$

$$\frac{f_a}{0.60 F_y} + \frac{f_{bx}}{F_{bx}} + \frac{f_{by}}{F_{by}} \leq 1$$

When $f_a / F_a \leq 0.15$, the following equation may be used instead of the preceding two:

$$\frac{f_a}{F_a} + \frac{f_{bx}}{F_{bx}} + \frac{f_{by}}{F_{by}} \leq 1$$

In the preceding equations, subscripts x and y indicate the axis of bending about which the stress occurs, and

F_a = axial stress that would be permitted if axial force alone existed, ksi (MPa)

F_b = compressive bending stress that would be permitted if bending moment alone existed, ksi (MPa)

F'_e = $149{,}000/(Kl_b/r_b)^2$, ksi (MPa); as for F_a, F_b, and $0.6F_y$, F'_e may be increased one-third for wind and seismic loads

l_b = actual unbraced length in plane of bending, in (mm)

r_b = radius of gyration about bending axis, in (mm)

K = effective-length factor in plane of bending

f_a = computed axial stress, ksi (MPa)

f_b = computed compressive bending stress at point under consideration, ksi (MPa)

C_m = adjustment coefficient

WEBS UNDER CONCENTRATED LOADS

Criteria for Buildings

The AISC specification for ASD for buildings places a limit on compressive stress in webs to prevent local web yielding. For a rolled beam, bearing stiffeners are required at a concentrated load if the stress f_a, ksi (MPa), at the toe of the web fillet exceeds $F_a = 0.66F_{yw}$, where F_{yw} is the minimum specified yield stress of the web steel, ksi (MPa). In the calculation of the stressed area, the load may be assumed distributed over the distance indicated in Fig. 9.4.

For a concentrated load applied at a distance larger than the depth of the beam from the end of the beam:

$$f_a = \frac{R}{t_w(N + 5k)}$$

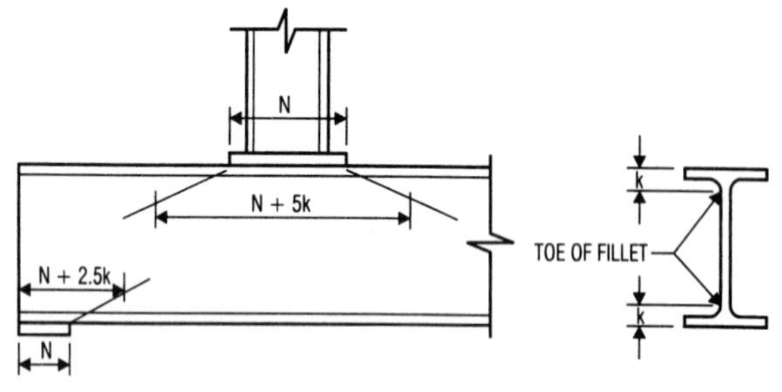

FIGURE 9.4 For investigating web yielding, stresses are assumed to be distributed over lengths of web indicated at the bearings, where N is the length of bearing plates, and k is the distance from outer surface of beam to the toe of the fillet.

where R = concentrated load of reaction, kip (kN)

t_w = web thickness, in (mm)

N = length of bearing, in (mm), (for end reaction, not less than k)

k = distance, in (mm), from outer face of flange to web toe of fillet

For a concentrated load applied close to the beam end:

$$f_a = \frac{R}{t_w(N + 2.5k)}$$

To prevent web crippling, the AISC specification requires that bearing stiffeners be provided on webs where concentrated loads occur when the compressive force exceeds R, kip (kN), computed from the following:

For a concentrated load applied at a distance from the beam end of at least $d/2$, where d is the depth of beam:

$$R = 67.5t_w^2 \left[1 + 3\left(\frac{N}{d}\right)\left(\frac{t_w}{t_f}\right)^{1.5}\right] \sqrt{F_{yw} t_f / t_w}$$

where t_f = flange thickness, in (mm)

For a concentrated load applied closer than $d/2$ from the beam end:

$$R = 34t_w^2 \left[1 + 3\left(\frac{N}{d}\right)\left(\frac{t_w}{t_f}\right)^{1.5}\right] \sqrt{F_{yw} t_f / t_w}$$

If stiffeners are provided and extend at least one-half of the web, R need not be computed.

Another consideration is prevention of sidesway web buckling. The AISC specification requires bearing stiffeners when the compressive force from a concentrated load exceeds limits that depend on the relative slenderness of web and flange r_{wf} and whether or not the loaded flange is restrained against rotation:

$$r_{wf} = \frac{d_c / t_w}{l / b_f}$$

where l = largest unbraced length, in (mm), along either top or bottom flange at point of application of load

b_f = flange width, in (mm)

d_c = web depth clear of fillets = $d - 2k$

Stiffeners are required if the concentrated load exceeds R, kip (kN), computed from

$$R = \frac{6800 t_w^3}{h} (1 + 0.4 r_{wf}^3)$$

where h = clear distance, in (mm), between flanges, and r_{wf} is less than 2.3 when the loaded flange is restrained against rotation. If the loaded flange is not restrained and r_{wf} is less than 1.7,

$$R = 0.4 r_{wf}^3 \frac{6800 t_w^3}{h}$$

R need not be computed for larger values of r_{wf}.

DESIGN OF STIFFENERS UNDER LOADS

AISC requires that fasteners or welds for end connections of beams, girders, and trusses be designed for the combined effect of forces resulting from moment and shear induced by the rigidity of the connection. When flanges or moment-connection plates for end connections of beams and girders are welded to the flange of an I- or H-shape column, a pair of column-web stiffeners having a combined cross-sectional area A_{st} not less than that calculated from the following equations must be provided whenever the calculated value of A_{st} is positive:

$$A_{st} = \frac{P_{bf} - F_{yc}t_{wc}(t_b + 5K)}{F_{yst}}$$

where F_{yc} = column yield stress, ksi (MPa)

F_{yst} = stiffener yield stress, ksi (MPa)

K = distance, in (mm), between outer face of column flange and web toe of its fillet, if column is rolled shape, or equivalent distance if column is welded shape

P_{bf} = computed force, kip (kN), delivered by flange of moment-connection plate multiplied by $5/3$, when computed force is due to live and dead load only, or by $4/3$, when computed force is due to live and dead load in conjunction with wind or earthquake forces

t_{wc} = thickness of column web, in (mm)

t_b = thickness of flange or moment-connection plate delivering concentrated force, in (mm)

Notwithstanding the preceding requirements, a stiffener or a pair of stiffeners must be provided opposite the beam-compression flange when the column-web depth clear of fillets d_c is greater than

$$d_c = \frac{4100 t_{wc}^3 \sqrt{F_{yc}}}{P_{bf}}$$

and a pair of stiffeners should be provided opposite the tension flange when the thickness of the column flange t_f is less than

$$t_f = 0.4 \sqrt{\frac{P_{bf}}{F_{yc}}}$$

Stiffeners required by the preceding equations should comply with the following additional criteria:

1. The width of each stiffener plus half the thickness of the column web should not be less than one-third the width of the flange or moment-connection plate delivering the concentrated force.
2. The thickness of stiffeners should not be less than $t_b/2$.
3. The weld-joining stiffeners to the column web must be sized to carry the force in the stiffener caused by unbalanced moments on opposite sides of the column.

FASTENERS IN BUILDINGS

The AISC specification for allowable stresses for buildings specifies allowable unit tension and shear stresses on the cross-sectional area on the unthreaded body area of bolts and threaded parts.

(Generally, rivets should not be used in direct tension.) When wind or seismic load are combined with gravity loads, the allowable stresses may be increased one-third.

Most building construction is done with bearing-type connections. Allowable bearing stresses apply to both bearing-type and slip-critical connections. In buildings, the allowable bearing stress F_p, ksi (MPa), on projected area of fasteners is

$$F_p = 1.2F_u$$

where F_u is the tensile strength of the connected part, ksi (MPa). Distance measured in the line of force to the nearest edge of the connected part (end distance) should be at least $1.5d$, where d is the fastener diameter. The center-to-center spacing of fasteners should be at least $3d$.

COMPOSITE CONSTRUCTION

In composite construction, steel beams and a concrete slab are connected so that they act together to resist the load on the beam. The slab, in effect, serves as a cover plate. As a result, a lighter steel section may be used.

Construction In Buildings

There are two basic methods of composite construction.

Method 1. The steel beam is entirely encased in the concrete. Composite action in this case depends on the steel-concrete bond alone. Because the beam is completely braced laterally, the allowable stress in the flanges is

$0.66F_y$, where F_y is the yield strength, ksi (MPa), of the steel. Assuming the steel to carry the full dead load and the composite section to carry the live load, the maximum unit stress, ksi (MPa), in the steel is

$$f_s = \frac{M_D}{S_s} + \frac{M_L}{S_{tr}} \leq 0.66F_y$$

where M_D = dead-load moment, in·kip (kN·mm)

M_L = live-load moment, in·kip (kN·mm)

S_s = section modulus of steel beam, in^3 (mm^3)

S_{tr} = section modulus of transformed composite section, in^3 (mm^3)

An alternative, shortcut method is permitted by the AISC specification. It assumes that the steel beam carries both live and dead loads and compensates for this by permitting a higher stress in the steel:

$$f_s = \frac{M_D + M_L}{S_s} \leq 0.76F_y$$

Method 2. The steel beam is connected to the concrete slab by shear connectors. Design is based on ultimate load and is independent of the use of temporary shores to support the steel until the concrete hardens. The maximum stress in the bottom flange is

$$f_s = \frac{M_D + M_L}{S_{tr}} \leq 0.66F_y$$

To obtain the transformed composite section, treat the concrete above the neutral axis as an equivalent steel area by dividing the concrete area by n, the ratio of modulus of elasticity of steel to that of the concrete. In determination of the transformed section, only a portion of the concrete slab over the beam may be considered effective in resisting compressive flexural stresses (positive-moment regions). The width of slab on either side of the beam centerline that may be considered effective should not exceed any of the following:

1. One-eighth of the beam span between centers of supports
2. Half the distance to the centerline of the adjacent beam
3. The distance from beam centerline to edge of slab (Fig. 9.5)

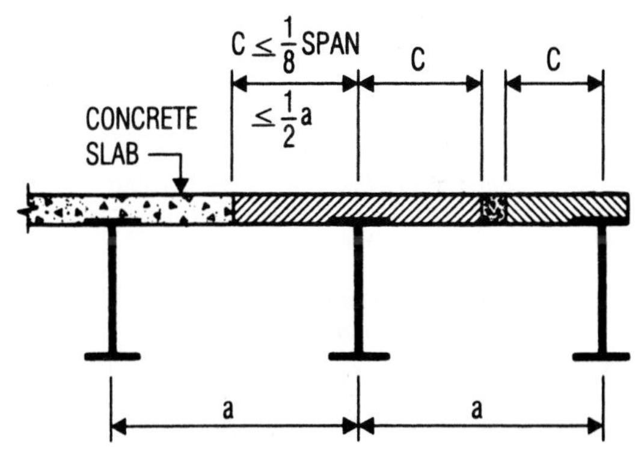

FIGURE 9.5 Limitations on effective width of concrete slab in a composite steel-concrete beam.

NUMBER OF CONNECTORS REQUIRED FOR BUILDING CONSTRUCTION

The total number of connectors to resist V_h is computed from V_h/q, where q is the allowable shear for one connector, kip (kN). Values of q for connectors in buildings are given in structural design guides.

The required number of shear connectors may be spaced uniformly between the sections of maximum and zero moment. Shear connectors should have at least 1 in (25.4 mm) of concrete cover in all directions; and unless studs are located directly over the web, stud diameters may not exceed 2.5 times the beam-flange thickness.

With heavy concentrated loads, the uniform spacing of shear connectors may not be sufficient between a concentrated load and the nearest point of zero moment. The number of shear connectors in this region should be at least

$$N_2 = \frac{N_1[(M\beta/M_{\max}) - 1]}{\beta - 1}$$

where M = moment at concentrated load, ft·kip (kN·m)

$M_{\max}$ = maximum moment in span, ft·kip (kN·m)

N_1 = number of shear connectors required between $M_{\max}$ and zero moment

$\beta = S_{tr}/S_s$ or S_{eff}/S_s, as applicable

S_{eff} = effective section modulus for partial composite action, in^3 (mm^3)

Shear on Connectors

The total horizontal shear to be resisted by the shear connectors in building construction is taken as the smaller of the values given by the following two equations:

$$V_h = \frac{0.85 f'_c A_c}{2}$$

$$V_h = \frac{A_s F_y}{2}$$

where V_h = total horizontal shear, kip (kN), between maximum positive moment and each end of steel beams (or between point of maximum positive moment and point of contraflexure in continuous beam)

f'_c = specified compressive strength of concrete at 28 days, ksi (MPa)

A_c = actual area of effective concrete flange, in^2 (mm^2)

A_s = area of steel beam, in^2 (mm^2)

In continuous composite construction, longitudinal reinforcing steel may be considered to act compositely with the steel beam in negative-moment regions. In this case, the total horizontal shear, kip (kN), between an interior support and each adjacent point of contraflexure should be taken as

$$V_h = \frac{A_{sr} F_{yr}}{2}$$

where A_{sr} = area of longitudinal reinforcement at support within effective area, in^2 (mm^2); and F_{yr} = specified minimum yield stress of longitudinal reinforcement, ksi (MPa).

PONDING CONSIDERATIONS IN BUILDINGS

Flat roofs on which water may accumulate may require analysis to ensure that they are stable under ponding conditions. A flat roof may be considered stable and an analysis does not need to be made if both of the following two equations are satisfied:

$$C_p + 0.9C_s \leq 0.25$$

$$I_d \geq 25S^4/10^6$$

where $C_p = 32L_sL_p^4/10^7I_p$

$C_s = 32SL_s^4/10^7I_s$

L_p = length, ft (m), of primary member or girder

L_s = length, ft (m), of secondary member or purlin

S = spacing, ft (m), of secondary members

I_p = moment of inertia of primary member, in^4 (mm^4)

I_s = moment of inertia of secondary member, in^4 (mm^4)

I_d = moment of inertia of steel deck supported on secondary members, in^4/ft (mm^4/m)

For trusses and other open-web members, I_s should be decreased 15 percent. The total bending stress due to dead loads, gravity live loads, and ponding should not exceed $0.80F_y$, where F_y is the minimum specified yield stress for the steel.

CHAPTER 10

BRIDGE AND SUSPENSION-CABLE FORMULAS

SHEAR STRENGTH DESIGN FOR BRIDGES

Based on the *American Association of State Highway and Transportation Officials* (AASHTO) specifications for *load-factor design* (LFD), the shear capacity, kip (kN), may be computed from

$$V_u = 0.58F_y h t_w C$$

for flexural members with unstiffened webs with $h/t_w < 150$ or for girders with stiffened webs but a/h exceeding 3 or $67{,}600(h/t_w)^2$:

$$C = 1.0 \qquad \text{when} \qquad \frac{h}{t_w} < \beta$$

$$= \frac{\beta}{h/t_w} \qquad \text{when} \qquad \beta \le \frac{h}{t_w} \le 1.25\beta$$

$$= \frac{45{,}000k}{F_y (h/t_w)^2} \qquad \text{when} \qquad \frac{h}{t_w} > 1.25\beta$$

For girders with transverse stiffeners and a/h less than 3 and $67{,}600(h/t_w)^2$, the shear capacity is given by

$$V_u = 0.58F_y d t_w \left[C + \frac{1 - C}{1.15\sqrt{1 + (a/h)^2}} \right]$$

Stiffeners are required when the shear exceeds V_u.

Chap. 9, "Building and Structures Formulas," for symbols used in the preceding equations.

ALLOWABLE-STRESS DESIGN FOR BRIDGE COLUMNS

In the AASHTO bridge-design specifications, allowable stresses in concentrically loaded columns are determined from the following equations:

When Kl/r is less than C_c,

$$F_a = \frac{F_y}{2.12}\left[1 - \frac{(Kl/r)^2}{2C_c^2}\right]$$

When Kl/r is equal to or greater than C_c,

$$F_a = \frac{\pi^2 E}{2.12(Kl/r)^2} = \frac{135{,}000}{(Kl/r)^2}$$

See Table 10.1.

TABLE 10.1 Column Formulas for Bridge Design

Yield Strength, ksi	(MPa)	C_c	Allowable Stress, ksi (MPa) $Kl/r < C_c$	Allowable Stress, ksi (MPa) $Kl/r \geq C_c$
36	(248)	126.1	$16.98 - 0.00053(Kl/r)^2$	
50	(345)	107.0	$23.58 - 0.00103(Kl/r)^2$	$\frac{135{,}000}{(Kl/r)^2}$
90	(620)	79.8	$42.45 - 0.00333(Kl/r)^2$	
100	(689)	75.7	$47.17 - 0.00412(Kl/r)^2$	

LOAD-AND-RESISTANCE FACTOR DESIGN FOR BRIDGE COLUMNS

Compression members designed by LFD should have a maximum strength, kip (kN),

$$P_u = 0.85 A_s F_{cr}$$

where A_s = gross effective area of column cross section, in^2 (mm^2).

For $KL_c/r \leq \sqrt{2\pi^2 E/F_y}$:

$$F_{cr} = F_y \left[1 - \frac{F_y}{4\pi^2 E} \left(\frac{KL_c}{r} \right)^2 \right]$$

For $KL_c/r > \sqrt{2\pi^2 E/F_y}$:

$$F_{cr} = \frac{\pi^2 E}{(KL_c/r)^2} = \frac{286{,}220}{(KL_c/r)^2}$$

where F_{cr} = buckling stress, ksi (MPa)

F_y = yield strength of the steel, ksi (MPa)

K = effective-length factor in plane of buckling

L_c = length of member between supports, in (mm)

r = radius of gyration in plane of buckling, in (mm)

E = modulus of elasticity of the steel, ksi (MPa)

The preceding equations can be simplified by introducing a Q factor:

$$Q = \left(\frac{KL_c}{r}\right)^2 \frac{F_y}{2\pi^2 E}$$

Then, the preceding equations can be rewritten as shown next:

For $Q \leq 1.0$:

$$F_{cr} = \left(1 - \frac{Q}{2}\right) F_y$$

For $Q > 1.0$:

$$F_{cr} = \frac{F_y}{2Q}$$

ALLOWABLE-STRESS DESIGN FOR BRIDGE BEAMS

AASHTO gives the allowable unit (tensile) stress in bending as $F_b = 0.55F_y$. The same stress is permitted for compression when the compression flange is supported laterally for its full length by embedment in concrete or by other means.

When the compression flange is partly supported or unsupported in a bridge, the allowable bending stress, ksi (MPa), is (Table 10.2):

TABLE 10.2 Allowable Bending Stress in Braced Bridge Beams†

F_y	F_b
36 (248)	20 (138)
50 (345)	27 (186)

† Units in ksi (MPa).

$$F_b = \left(\frac{5 \times 10^7 C_b}{S_{xc}}\right)\left(\frac{I_{yc}}{L}\right) \times \sqrt{\frac{0.772J}{I_{yc}} + 9.87\left(\frac{d}{L}\right)^2} \leq 0.55F_y$$

where L = length, in (mm), of unsupported flange between connections of lateral supports, including knee braces

S_{xc} = section modulus, in^3 (mm^3), with respect to the compression flange

I_{yc} = moment of inertia, in^4 (mm^4) of the compression flange about the vertical axis in the plane of the web

$J = 1/3(b_c t_c^3 + b_t t_c^3 + D t_w^3)$

b_c = width, in (mm), of compression flange

b_t = width, in (mm), of tension flange

t_c = thickness, in (mm), of compression flange

t_t = thickness, in (mm), of tension flange

t_w = thickness, in (mm), of web

D = depth, in (mm), of web

d = depth, in (mm), of flexural member

In general, the moment-gradient factor C_b may be computed from the next equation. It should be taken as unity, however, for unbraced cantilevers and members in which the moment within a significant portion of the unbraced length is equal to, or greater than, the larger of the segment end moments. If cover plates are used, the allowable static stress at the point of cutoff should be computed from the preceding equation.

The moment-gradient factor may be computed from

$$C_b = 1.75 + 1.05\frac{M_1}{M_2} + 0.3\left(\frac{M_1}{M_2}\right)^2 \leq 2.3$$

where M_1 = smaller beam end moment, and M_2 = larger beam end moment. The algebraic sign of M_1/M_2 is positive for double-curvature bending and negative for single-curvature bending.

STIFFENERS ON BRIDGE GIRDERS

The minimum moment of inertia, in^4 (mm^4), of a transverse stiffener should be at least

$$I = a_o t^3 J$$

where $J = 2.5h^2/a_o^2 - 2 \geq 0.5$

h = clear distance between flanges, in (mm)

a_o = actual stiffener spacing, in (mm)

t = web thickness, in (mm)

For paired stiffeners, the moment of inertia should be taken about the centerline of the web; for single stiffeners, about the face in contact with the web.

The gross cross-sectional area of intermediate stiffeners should be at least

$$A = \left[0.15BDt_w(1 - C)\frac{V}{V_u} - 18t_w^2\right]Y$$

where Y is the ratio of web-plate yield strength to stiffener-plate yield strength, $B = 1.0$ for stiffener pairs, 1.8 for single angles, and 2.4 for single plates; and C is defined in the earlier section, "Allowable-Stress Design for Bridge Columns." V_u should be computed from the previous section equations in "Shear Strength Design for Bridges."

The width of an intermediate transverse stiffener, plate, or outstanding leg of an angle should be at least 2 in (50.8 mm), plus $^1/_{30}$ of the depth of the girder and preferably not less than $^1/_4$ of the width of the flange. Minimum thickness is $^1/_{16}$ of the width.

Longitudinal Stiffeners

These should be placed with the center of gravity of the fasteners $h/5$ from the toe, or inner face, of the compression flange. Moment of inertia, in^4 (mm^4), should be at least

$$I = ht^3\left(2.4\frac{a_o^2}{h^2} - 0.13\right)$$

where a_o = actual distance between transverse stiffeners, in (mm); and t = web thickness, in (mm).

Thickness of stiffener, in (mm), should be at least $b\sqrt{f_b}/71.2$, where b is the stiffener width, in (mm), and f_b is the flange compressive bending stress, ksi (MPa). The bending stress in the stiffener should not exceed that allowable for the material.

HYBRID BRIDGE GIRDERS

These may have flanges with larger yield strength than the web and may be composite or noncomposite with a concrete slab, or they may utilize an orthotropic-plate deck as the top flange.

Computation of bending stresses and allowable stresses is generally the same as that for girders with uniform yield strength. The bending stress in the web, however, may exceed the allowable bending stress if the computed flange bending stress does not exceed the allowable stress multiplied by

$$R = 1 - \frac{\beta\psi(1-\alpha)^2(3-\psi+\psi\alpha)}{6+\beta\psi(3-\psi)}$$

where α = ratio of web yield strength to flange yield strength

ψ = distance from outer edge of tension flange or bottom flange of orthotropic deck to neutral axis divided by depth of steel section

β = ratio of web area to area of tension flange or bottom flange of orthotropic-plate bridge

LOAD-FACTOR DESIGN FOR BRIDGE BEAMS

For LFD of symmetrical beams, there are three general types of members to consider: compact, braced noncompact, and unbraced sections. The maximum strength of each (moment, in·kip) (mm·kN) depends on member dimensions and unbraced length, as well as on applied shear and axial load (Table 10.3).

The maximum strengths given by the formulas in Table 10.3 apply only when the maximum axial stress does not exceed $0.15F_yA$, where A is the area of the member. Symbols used in Table 10.3 are defined as follows:

F_y = steel yield strength, ksi (MPa)

Z = plastic section modulus, in^3 (mm^3)

S = section modulus, in^3 (mm^3)

b' = width of projection of flange, in (mm)

d = depth of section, in (mm)

h = unsupported distance between flanges, in (mm)

M_1 = smaller moment, in·kip (mm·kN), at ends of unbraced length of member

$M_u = F_y Z$

M_1/M_u is positive for single-curvature bending.

TABLE 10.3 Design Criteria for Symmetrical Flexural Sections for Load-Factor Design of Bridges

Type of section	Maximum bending strength M_u, in·kip (mm·kN)	Flange minimum thickness t_f, in (mm)	Web minimum thickness t_u, in (mm)	Maximum unbraced length l_b, in (mm)
Compact†	$F_y Z$	$\frac{b'\sqrt{F_y}}{65.0}$	$\frac{d\sqrt{F_y}}{608}$	$\frac{[3600 - 2200(M_1/M_u)]r_y}{F_y}$
Braced noncompact†	$F_y S$	$\frac{b'\sqrt{F_y}}{69.6}$	$\frac{h}{150}$	$\frac{20{,}000\, A_f}{F_y d}$
Unbraced	See AASHTO specification			

† Straight-line interpolation between compact and braced noncompact moments may be used for intermediate criteria, except that $t_w \leq d\sqrt{F_y}/608$ should be maintained as well as the following: For compact sections, when both b'/t_f and d/t_w exceed 75% of the limits for these ratios, the following interaction equation applies:

$$\frac{d}{t_w} + 9.35\frac{b'}{t_f} \leq \frac{1064}{\sqrt{F_{yf}}}$$

where F_{yf} is the yield strength of the flange, ksi (MPa); t_w is the web thickness, in (mm); and t_f = flange thickness, in (mm).

BEARING ON MILLED SURFACES

For highway design, AASHTO limits the allowable bearing stress on milled stiffeners and other steel parts in contact to $F_p = 0.80F_u$. Allowable bearing stresses on pins are given in Table 10.4.

The allowable bearing stress for expansion rollers and rockers used in bridges depends on the yield point in tension F_y of the steel in the roller or the base, whichever is smaller. For diameters up to 25 in (635 mm) the allowable stress, kip/linear in (kN/mm), is

$$p = \frac{F_y - 13}{20} 0.6d$$

For diameters from 25 to 125 in (635 to 3175 mm),

$$p = \frac{F_y - 13}{20} 3\sqrt{d}$$

where d = diameter of roller or rocker, in (mm).

TABLE 10.4 Allowable Bearing Stresses on Pins[†]

		Bridges	
F_y	Buildings $F_p = 0.90F_y$	Pins subject to rotation $F_p = 0.40F_y$	Pins not subject to rotation $F_p = 0.80F_y$
36 (248)	33 (227)	14 (96)	29 (199)
50 (344)	45 (310)	20 (137)	40 (225)

[†]Units in ksi (MPa).

BRIDGE FASTENERS

For bridges, AASHTO specifies the working stresses for bolts. Bearing-type connections with high-strength bolts are limited to members in compression and secondary members. The allowable bearing stress is

$$F_p = 1.35F_u$$

where F_p = allowable bearing stress, ksi (MPa); and F_u = tensile strength of the connected part, ksi (MPa) (or as limited by allowable bearing on the fasteners). The allowable bearing stress on A307 bolts is 20 ksi (137.8 MPa) and on structural-steel rivets is 40 ksi (275.6 MPa).

COMPOSITE CONSTRUCTION IN HIGHWAY BRIDGES

Shear connectors between a steel girder and a concrete slab in composite construction in a highway bridge should be capable of resisting both horizontal and vertical movement between the concrete and steel. Maximum spacing for shear connectors generally is 24 in (609.6 mm), but wider spacing may be used over interior supports, to avoid highly stressed portions of the tension flange (Fig. 10.1). Clear depth of concrete cover over shear connectors should be at least 2 in (50.8 mm), and they should extend at least 2 in (50.8 mm) above the bottom of the slab.

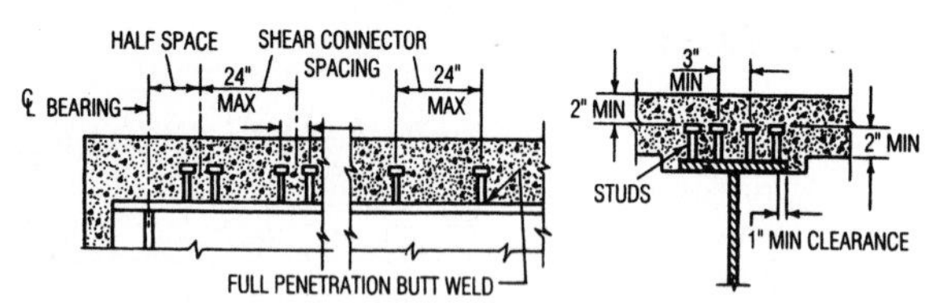

FIGURE 10.1 Maximum pitch for stud shear connectors in composite beams: 1 in (25.4 mm), 2 in (50.8 mm), 3 in (76.2 mm), and 24 in (609.6 mm).

Span/Depth Ratios

In bridges, for composite beams, preferably the ratio of span/steel beam depth should not exceed 30 and the ratio of span/depth of steel beam plus slab should not exceed 25.

Effective Width of Slabs

For a composite interior girder, the effective width assumed for the concrete flange should not exceed any of the following:

1. One-fourth the beam span between centers of supports
2. Distance between centerlines of adjacent girders
3. Twelve times the least thickness of the slab

For a girder with the slab on only one side, the effective width of slab should not exceed any of the following:

1. One-twelfth the beam span between centers of supports
2. Half the distance to the centerline of the adjacent girder
3. Six times the least thickness of the slab

Bending Stresses

In composite beams in bridges, stresses depend on whether or not the members are shored; they are determined as for beams in buildings (see "Composite Construction" in Chap. 9,"Building and Structures Formulas"), except that the stresses in the steel may not exceed $0.55F_y$. (See the following equations.)

For unshored members:

$$f_s = \frac{M_D}{S_s} + \frac{M_L}{S_{tr}} \leq 0.55F_y$$

where f_y = yield strength, ksi (MPa).

For shored members:

$$f_s = \frac{M_D + M_L}{S_{tr}} \leq 0.55F_y$$

where f_s = stress in steel, ksi (MPa)

M_D = dead-load moment, in·kip (kN·mm)

M_L = live-load moment, in·kip (kN·mm)

S_s = section modulus of steel beam, in^3 (mm^3)

S_{tr} = section modulus of transformed composite section, in^3 (mm^3)

V_r = shear range (difference between minimum and maximum shears at the point) due to live load and impact, kip (kN)

Q = static moment of transformed compressive concrete area about neutral axis of transformed section, in^3 (mm^3)

I = moment of inertia of transformed section, in^4 (mm^4)

Shear Range

Shear connectors in bridges are designed for fatigue and then are checked for ultimate strength. The horizontal-shear range for fatigue is computed from

$$S_r = \frac{V_r Q}{I}$$

where S_r = horizontal-shear range at the juncture of slab and beam at point under consideration, kip/linear in (kN/linear mm).

The transformed area is the actual concrete area divided by n (Table 10.5).

The allowable range of horizontal shear Z_r, kip (kN), for an individual connector is given by the next two equations, depending on the connector used.

TABLE 10.5 Ratio of Moduli of Elasticity of Steel and Concrete for Bridges

f'_c for concrete	$n = \frac{E_s}{E_c}$
2.0–2.3	11
2.4–2.8	10
2.9–3.5	9
3.6–4.5	8
4.6–5.9	7
6.0 and over	6

For channels, with a minimum of $^3/_{16}$-in (4.76-mm) fillet welds along heel and toe:

$$Z_r = Bw$$

where w = channel length, in (mm), in transverse direction on girder flange; and

B = cyclic variable = 4.0 for 100,000 cycles, 3.0 for 500,000 cycles, 2.4 for 2 million cycles, and 2.1 for over 2 million cycles.

For welded studs (with height/diameter ratio $H/d \geq 4$):

$$Z_r = \alpha d^2$$

where d = stud diameter, in (mm); and

α = cyclic variable = 13.0 for 100,000 cycles, 10.6 for 500,000 cycles, 7.85 for 2 million cycles, and 5.5 for over 2 million cycles.

Required pitch of shear connectors is determined by dividing the allowable range of horizontal shear of all connectors at one section Z_r, kip (kN), by the horizontal range of shear S_r, kip per linear in (kN per linear mm).

NUMBER OF CONNECTORS IN BRIDGES

The ultimate strength of the shear connectors is checked by computation of the number of connectors required from

$$N = \frac{P}{\phi S_u}$$

where N = number of shear connectors between maximum positive moment and end supports

S_u = ultimate shear connector strength, kip (kN) [see Eqs. (10.1) and (10.2) that follow and AASHTO data]

ϕ = reduction factor = 0.85

P = force in slab, kip (kN)

At points of maximum positive moments, P is the smaller of P_1 and P_2, computed from

$$P_1 = A_s F_y$$

$$P_2 = 0.85 f_c' A_c$$

where A_c = effective concrete area, in^2 (mm^2)

f_c' = 28-day compressive strength of concrete, ksi (MPa)

A_s = total area of steel section, in^2 (mm^2)

F_y = steel yield strength, ksi (MPa)

The number of connectors required between points of maximum positive moment and points of adjacent maximum negative moment should equal or exceed N_2, given by

$$N_2 = \frac{P + P_3}{\phi S_u}$$

At points of maximum negative moments, the force in the slab P_3, is computed from

$$P_3 = A_{sr} F_{yr}$$

where A_{sr} = area of longitudinal reinforcing within effective flange, in^2 (mm^2); and F_{yr} = reinforcing steel yield strength, ksi (MPa).

Ultimate Shear Strength of Connectors in Bridges

For channels:

$$S_u = 17.4 \left(h + \frac{t}{2} \right) w \sqrt{f_c'} \qquad (10.1)$$

where h = average channel-flange thickness, in (mm)

t = channel-web thickness, in (mm)

w = channel length, in (mm)

For welded studs ($H/d \geq 4$ in (101.6 mm):

$$S_u = 0.4 d^2 \sqrt{f_c' E_c} \qquad (10.2)$$

ALLOWABLE-STRESS DESIGN FOR SHEAR IN BRIDGES

Based on the AASHTO specification for highway bridges, the allowable shear stress, ksi (MPa), may be computed from

$$F_v = \frac{F_y}{3} C \leq \frac{F_y}{3}$$

for flexural members with unstiffened webs with $h/t_w < 150$ or for girders with stiffened webs with a/h exceeding 3 and $67{,}600(h/t_w)^2$:

$$C = 1.0 \qquad \text{when} \qquad \frac{h}{t_w} \leq \beta$$

$$= \frac{\beta}{h/t_w} \qquad \text{when} \qquad \beta < \frac{h}{t_w} \leq 1.25\beta$$

$$= \frac{45{,}000k}{F_y (h/t_w)^2} \qquad \text{when} \qquad \frac{h}{t_w} > 1.25\beta$$

$$k = 5 \text{ if } \frac{a}{h} \text{ exceeds 3 or } 67{,}600 \left(\frac{h}{t_w}\right)^2$$

or stiffeners are not required

$$= 5 + \frac{5}{(a/h)^2} \qquad \text{otherwise}$$

$$\beta = 190 \sqrt{\frac{k}{F_y}}$$

For girders with transverse stiffeners and a/h less than 3 and $67{,}600(h/t_w)^2$, the allowable shear stress is given by

$$F_v = \frac{F_y}{3} \left[C + \frac{1 - C}{1.15 \sqrt{1 + (a/h)^2}} \right]$$

Stiffeners are required when the shear exceeds F_v.

MAXIMUM WIDTH/THICKNESS RATIOS FOR COMPRESSION ELEMENTS FOR HIGHWAY BRIDGES

Table 10.6 gives a number of formulas for maximum width/thickness ratios for compression elements for highway bridges. These formulas are valuable for highway bridge design.

SUSPENSION CABLES

Parabolic Cable Tension and Length

Steel cables are often used in suspension bridges to support the horizontal roadway load (Fig. 10.2). With a uniformly distributed load along the horizontal, the cable assumes the form of a parabolic arc. Then, the tension at midspan is

$$H = \frac{wL^2}{8d}$$

where H = midspan tension, kip (N)

w = load on a unit horizontal distance, klf (kN/m)

L = span, ft (m)

d = sag, ft (m)

The tension at the supports of the cable is given by

$$T = \left[H^2 + \left(\frac{wL}{2}\right)^2\right]^{0.5}$$

TABLE 10.6 Maximum Width/Thickness Ratios b/t^a for Compression Elements for Highway Bridges[b]

Load-and-resistance-factor design[c]		
Description of element	Compact	Noncompact[d]
Flange projection of rolled or fabricated I-shaped beams	$\frac{65}{\sqrt{F_y}}$	$\frac{70^e}{\sqrt{F_y}}$
Webs in flexural compression	$\frac{608}{\sqrt{F_y}}$	150

Allowable-stress design[f]			
		$f_a = 0.44F_y$	
Description of element	$f_a < 0.44F_y$	F_y = 36 ksi (248 MPa)	F_y = 50 ksi (344.5 MPa)
Plates supported in one side and outstanding legs of angles			
In main members	$\frac{51}{\sqrt{f_a}} \le 12$	12	11

In bracing and other secondary members	$\frac{51}{\sqrt{f_a}} \leq 16$	12	11
Plates supported on two edges or webs of box shapes[g]	$\frac{126}{\sqrt{f_a}} \leq 45$	32	27
Solid cover plates supported on two edges or solid webs[h]	$\frac{158}{\sqrt{f_a}} \leq 50$	40	34
Perforated cover plates supported on two edges for box shapes	$\frac{190}{\sqrt{f_a}} \leq 55$	48	41

[a] b = width of element or projection; t = thickness. The point of support is the inner line of fasteners or fillet welds connecting a plate to the main segment or the root of the flange of rolled shapes. In LRFD, for webs of compact sections, $b = d$, the beam depth; and for noncompact sections, $b = D$, the unsupported distance between flange components.
[b] As required in AASHTO "Standard Specification for Highway Bridges." The specifications also provide special limitations on plate-girder elements.
[c] F_y = specified minimum yield stress, ksi (MPa), of the steel.
[d] Elements with width/thickness ratios that exceed the noncompact limits should be designed as slender elements.
[e] When the maximum bending moment M is less than the bending strength M_u, b/t in the table may be multiplied by $\sqrt{M_u/M}$.
[f] f_a = computed axial compression stress, ksi (MPa).
[g] For box shapes consisting of main plates, rolled sections, or component segments with cover plates.
[h] For webs connecting main members or segments for H or box shapes.

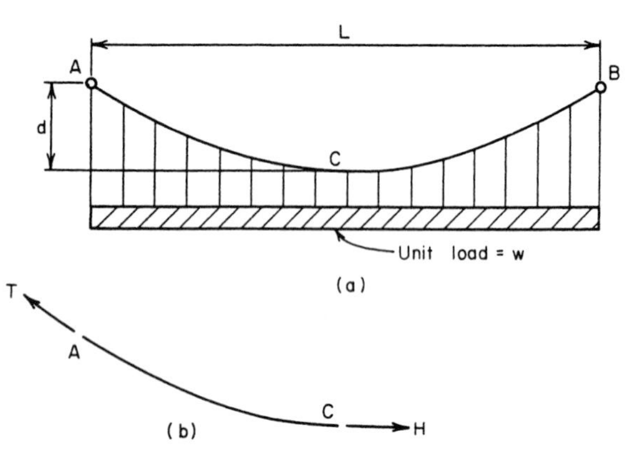

FIGURE 10.2 Cable supporting load uniformly distributed along the horizontal.

where T = tension at supports, kip (N); and other symbols are as before.

Length of the cable, S, when d/L is 1/20, or less, can be approximated from

$$S = L + \frac{8d^2}{3L}$$

where S = cable length, ft (m).

Catenary Cable Sag and Distance between Supports

A cable of uniform cross section carrying only its own weight assumes the form of a catenary. By using the same previous notation, the catenary parameter, c, is found from

$$d + c = \frac{T}{w}$$

Then

$$c = \left[(d + c)^2 - \left(\frac{S}{2} \right)^2 \right]^{0.5}$$

$$\text{Sag} = d + c \quad \text{ft (m)}$$

Span length then is $L = 2c$, with the previous same symbols.

GENERAL RELATIONS FOR SUSPENSION CABLES

Catenary

For any simple cable (Fig. 10.3) with a load of q_o per unit length of cable, kip/ft (N/m), the catenary length s, ft (m), measured from the low point of the cable is, with symbols as given in Fig. 10.3, ft (m),

$$s = \frac{H}{q_o} \sinh \frac{q_o x}{H} = x + \frac{1}{3!} \left(\frac{q_o}{H} \right)^2 x^3 + \cdots$$

Tension at any point is

$$T = \sqrt{H^2 + q_o^2 s^2} = H + q_o y$$

The distance from the low point C to the left support is

$$a = \frac{H}{q_o} \cosh^{-1} \left(\frac{q_o}{H} f_L + 1 \right)$$

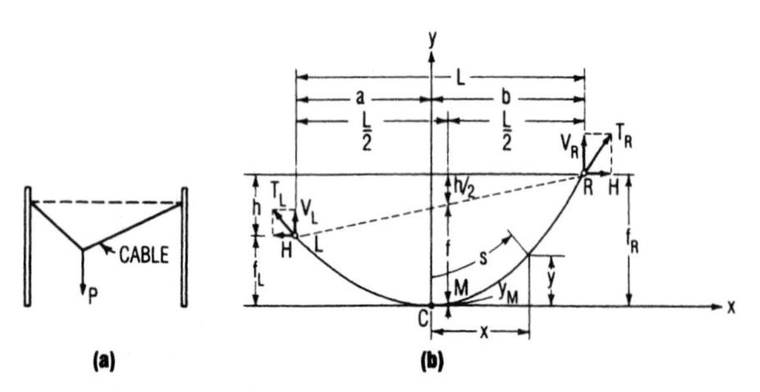

FIGURE 10.3 Simple cables. (*a*) Shape of cable with concentrated load; (*b*) shape of cable with supports at different levels.

where f_L = vertical distance from C to L, ft (m). The distance from C to the right support R is

$$b = \frac{H}{q_o} \cosh^{-1} \left(\frac{q_o}{H} f_R + 1 \right)$$

where f_R = vertical distance from C to R.

Given the sags of a catenary f_L and f_R under a distributed vertical load q_o, the horizontal component of cable tension H may be computed from

$$\frac{q_o l}{H} \cosh^{-1} \left(\frac{q_o f_L}{H} + 1 \right) + \cosh^{-1} \left(\frac{q_o f_R}{H} + 1 \right)$$

where l = span, or horizontal distance between supports L and $R = a + b$. This equation usually is solved by trial. A first estimate of H for substitution in the right-hand side of the equation may be obtained by approximating the catenary by a parabola. Vertical components of the reactions at the supports can be computed from

$$R_L = H \sinh \frac{q_o a}{H} \qquad R_R = H \sinh \frac{q_o b}{H}$$

Parabola

Uniform vertical live loads and uniform vertical dead loads other than cable weight generally may be treated as distributed uniformly over the horizontal projection of the cable. Under such loadings, a cable takes the shape of a parabola.

Take the origin of coordinates at the low point C (Fig. 10.3). If w_o is the load per foot (per meter) horizontally, the parabolic equation for the cable slope is

$$y = \frac{w_o x^2}{2H}$$

The distance from the low point C to the left support L is

$$a = \frac{l}{2} - \frac{Hh}{w_o l}$$

where l = span, or horizontal distance between supports L and $R = a + b$; h = vertical distance between supports.

The distance from the low point C to the right support R is

$$b = \frac{1}{2} + \frac{Hh}{w_o l}$$

Supports at Different Levels

The horizontal component of cable tension H may be computed from

$$H = \frac{w_o l^2}{h^2}\left(f_R - \frac{h}{2} \pm \sqrt{f_L f_R}\right) = \frac{w_o l^2}{8f}$$

where f_L = vertical distance from C to L

f_R = vertical distance from C to R

f = sag of cable measured vertically from chord LR midway between supports (at $x = Hh/w_o l$)

As indicated in Fig. 10.3*b*:

$$f = f_L + \frac{h}{2} - y_M$$

where $y_M = Hh^2/2w_o l^2$. The minus sign should be used when low point C is between supports. If the vertex of the parabola is not between L and R, the plus sign should be used.

The vertical components of the reactions at the supports can be computed from

$$V_L = w_o a = \frac{w_o l}{2} - \frac{Hh}{l}$$

$$V_r = w_o b = \frac{w_o l}{2} + \frac{Hh}{l}$$

Tension at any point is

$$T = \sqrt{H^2 + w_o^2 x^2}$$

Length of parabolic arc RC is

$$L_{RC} = \frac{b}{2}\sqrt{1 + \left(\frac{w_o b}{H}\right)^2} + \frac{H}{2w_o}\sinh\frac{w_o b}{H}$$

$$= b + \frac{1}{6}\left(\frac{w_o}{H}\right)^2 b^3 + \cdots$$

Length of parabolic arc LC is

$$L_{LC} = \frac{a}{2}\sqrt{1 + \left(\frac{w_o a}{H}\right)^2} + \frac{H}{2w_o}\sinh\frac{w_o a}{H}$$

$$= a + \frac{1}{6}\left(\frac{w_o}{H}\right)^2 a^3 + \cdots$$

Supports at Same Level

In this case, $f_L = f_R = f$, $h = 0$, and $a = b = l/2$. The horizontal component of cable tension H may be computed from

$$H = \frac{w_o l^2}{8f}$$

The vertical components of the reactions at the supports are

$$V_L = V_R = \frac{w_o l}{2}$$

Maximum tension occurs at the supports and equals

$$T_L = T_R = \frac{w_o l}{2}\sqrt{1 + \frac{l^2}{16f^2}}$$

Length of cable between supports is

$$L = \frac{1}{2}\sqrt{1 + \left(\frac{w_o l}{2H}\right)^2} + \frac{H}{w_o}\sinh\frac{w_o l}{2H}$$

$$= l\left(1 + \frac{8}{3}\frac{f^2}{l^2} - \frac{32}{5}\frac{f^4}{l^4} + \frac{256}{7}\frac{f^6}{l^6} + \cdots\right)$$

If additional uniformly distributed load is applied to a parabolic cable, the change in sag is approximately

$$\Delta f = \frac{15}{16}\frac{l}{f}\frac{\Delta L}{5 - 24f^2/l^2}$$

For a rise in temperature t, the change in sag is about

$$\Delta f = \frac{15}{16}\frac{l^2 ct}{f(5 - 24f^2/l^2)}\left(1 + \frac{8}{3}\frac{f^2}{l^2}\right)$$

where c = coefficient of thermal expansion.

Elastic elongation of a parabolic cable is approximately

$$\Delta L = \frac{Hl}{AE}\left(1 + \frac{16}{3}\frac{f^2}{l^2}\right)$$

where A = cross-sectional area of cable

E = modulus of elasticity of cable steel

H = horizontal component of tension in cable

If the corresponding change in sag is small, so that the effect on H is negligible, this change may be computed from

$$\Delta f = \frac{15}{16} \frac{Hl^2}{AEf} \frac{1 + 16f^2/3l^2}{5 - 24f^2/l^2}$$

For the general case of vertical dead load on a cable, the initial shape of the cable is given by

$$y_D = \frac{M_D}{H_D}$$

where M_D = dead-load bending moment that would be produced by load in a simple beam; and H_D = horizontal component of tension due to dead load.

For the general case of vertical live load on the cable, the final shape of the cable is given by

$$y_D + \delta = \frac{M_D + M_L}{H_D + H_L}$$

where δ = vertical deflection of cable due to live load

M_L = live-load bending moment that would be produced by live load in simple beam

H_L = increment in horizontal component of tension due to live load

Subtraction yields

$$\delta = \frac{M_L - H_L y_D}{H_D + H_L}$$

If the cable is assumed to take a parabolic shape, a close approximation to H_L may be obtained from

$$\frac{H_L}{AE} K = \frac{w_D}{H_D} \int_0^1 \delta \, dx - \frac{1}{2} \int_0^1 \delta'' \, \delta \, dx$$

$$K = l \left[\frac{1}{4} \left(\frac{5}{2} + \frac{16 f^2}{l^2} \right) \sqrt{1 + \frac{16 f^2}{l^2}} + \frac{3l}{32 f} \log_e \left(\frac{4 f}{l} + \sqrt{1 + \frac{16 f^2}{l^2}} \right) \right]$$

where $\delta'' = d^2\delta/dx^2$.

If elastic elongation and δ'' can be ignored,

$$H_L = \frac{\int_0^1 M_L \, dx}{\int_0^1 y_D \, dx} = \frac{3}{2 f l} \int_0^1 M_L \, dx$$

Thus, for a load uniformly distributed horizontally w_L,

$$\int_0^1 M_L \, dx = \frac{w_L l^3}{12}$$

and the increase in the horizontal component of tension due to live load is

$$H_L = \frac{3}{2 f l} \frac{w_L l^3}{12} = \frac{w_L l^2}{8 f} = \frac{w_L l^2}{8} \frac{8 H_D}{w_D l^2}$$

$$= \frac{w_L}{w_D} H_D$$

CABLE SYSTEMS

The cable that is concave downward (Fig. 10.4) usually is considered the load-carrying cable. If prestress in that cable exceeds that in the other cable, the natural frequencies of vibration of both cables always differ for any value of live load. To avoid resonance, the difference between the frequencies of the cables should increase with increase in load. Thus, the two cables tend to assume different shapes under specific dynamic loads. As a consequence, the resulting flow of energy from one cable to the other dampens the vibrations of both cables.

Natural frequency, cycles per second, of each cable may be estimated from

$$\omega_n = \frac{n\pi}{l}\sqrt{\frac{Tg}{w}}$$

where n = integer, 1 for fundamental mode of vibration, 2 for second mode, . . .

l = span of cable, ft (m)

w = load on cable, kip/ft (kN/m)

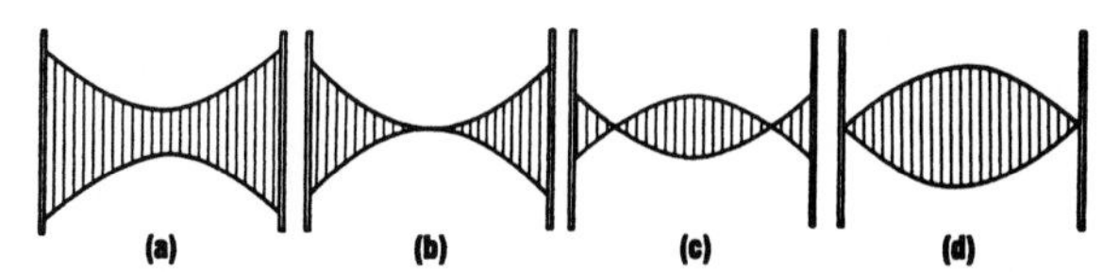

FIGURE 10.4 Planar cable systems: (*a*) completely separated cables; (*b*) cables intersecting at midspan; (*c*) crossing cables; (*d*) cables meeting at supports.

g = acceleration due to gravity = 32.2 ft/s^2

T = cable tension, kip (N)

The spreaders of a cable truss impose the condition that under a given load the change in sag of the cables must be equal. Nevertheless, the changes in tension of the two cables may not be equal. If the ratio of sag to span f/l is small (less than about 0.1), for a parabolic cable, the change in tension is given approximately by

$$\Delta H = \frac{16}{3} \frac{AEf}{l^2} \Delta f$$

where Δf = change in sag

A = cross-sectional area of cable

E = modulus of elasticity of cable steel

CHAPTER 11

HIGHWAY AND ROAD FORMULAS

CIRCULAR CURVES

Circular curves are the most common type of horizontal curve used to connect intersecting tangent (or straight) sections of highways or railroads. In most countries, two methods of defining circular curves are in use: the first, in general use in railroad work, defines the degree of curve as the central angle subtended by a *chord* of 100 ft (30.48 m) in length; the second, used in highway work, defines the degree of curve as the central angle subtended by an *arc* of 100 ft (30.48 m) in length.

The terms and symbols generally used in reference to circular curves are listed next and shown in Figs. 11.1 and 11.2.

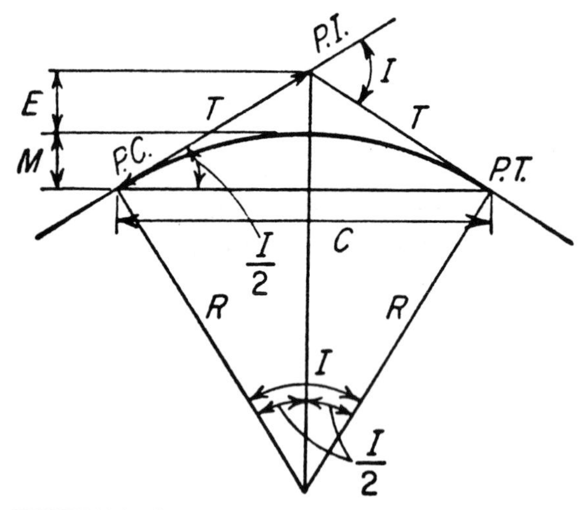

FIGURE 11.1 Circular curve.

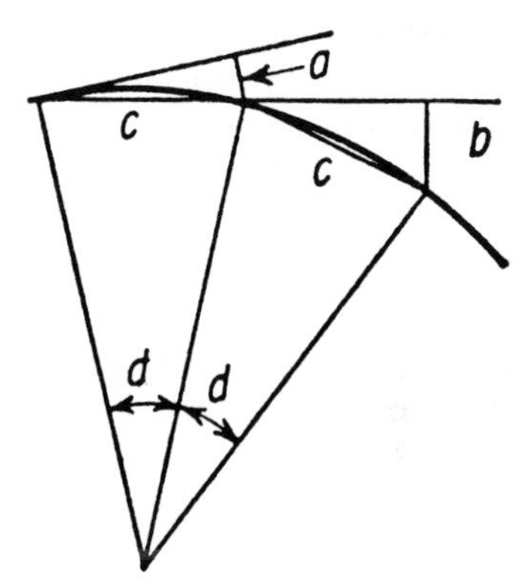

FIGURE 11.2 Offsets to circular curve.

PC = point of curvature, beginning of curve

PI = point of intersection of tangents

PT = point of tangency, end of curve

R = radius of curve, ft (m)

D = degree of curve (see previous text)

I = deflection angle between tangents at PI, also central angle of curve

T = tangent distance, distance from PI to PC or PT, ft (m)

L = length of curve from PC to PT measured on 100-ft (30.48-m) chord for chord definition, on arc for arc definition, ft (m)

C = length of long chord from PC to PT, ft (m)

E = external distance, distance from PI to midpoint of curve, ft (m)

M = midordinate, distance from midpoint of curve to midpoint of long chord, ft (m)

d = central angle for portion of curve ($d < D$)

l = length of curve (arc) determined by central angle d, ft (m)

c = length of curve (chord) determined by central angle d, ft (m)

a = tangent offset for chord of length c, ft (m)

b = chord offset for chord of length c, ft (m)

Equations of Circular Curves

$$R = \frac{5{,}729.578}{D}$$ exact for arc definition, approximate for chord definition

$$= \frac{50}{\sin \frac{1}{2} D}$$ exact for chord definition

$$T = R \tan \tfrac{1}{2} I$$ exact

$$E = R \operatorname{exsec} \tfrac{1}{2} I = R\,(\sec \tfrac{1}{2} I - 1)$$ exact

$$M = R \operatorname{vers} \tfrac{1}{2} I = R\,(1 - \cos \tfrac{1}{2} I)$$ exact

$$C = 2R \sin \tfrac{1}{2} I$$ exact

$$L = \frac{100I}{D}$$ exact

$$L - C = \frac{L^3}{24R^2} = \frac{C^3}{24R^2}$$ approximate

$$d = \frac{Dl}{100}$$ exact for arc definition

$$= \frac{Dc}{100} \qquad \text{approximate for chord definition}$$

$$\sin\frac{d}{z} = \frac{c}{2R} \qquad \text{exact for chord definition}$$

$$a = \frac{c^2}{2R} \qquad \text{approximate}$$

$$b = \frac{c^2}{R} \qquad \text{approximate}$$

PARABOLIC CURVES

Parabolic curves are used to connect sections of highways or railroads of differing gradient. The use of a parabolic curve provides a gradual change in direction along the curve. The terms and symbols generally used in reference to parabolic curves are listed next and are shown in Fig. 11.3.

PVC = point of vertical curvature, beginning of curve
PVI = point of vertical intersection of grades on either side of curve
PVT = point of vertical tangency, end of curve
G_1 = grade at beginning of curve, ft/ft (m/m)
G_2 = grade at end of curve, ft/ft (m/m)
L = length of curve, ft (m)
R = rate of change of grade, ft/ft^2 (m/m^2)
V = elevation of PVI, ft (m)

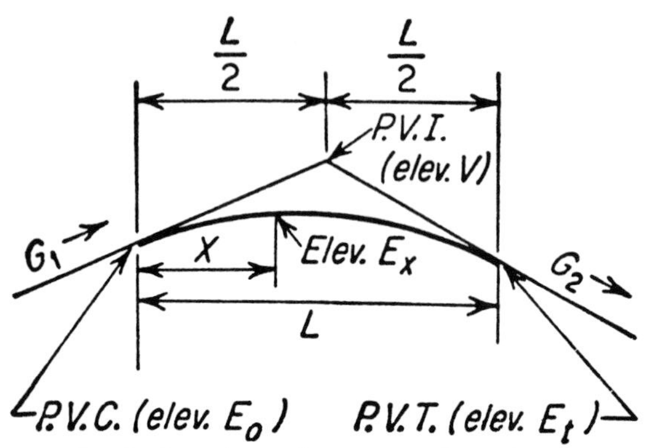

FIGURE 11.3 Vertical parabolic curve (summit curve).

E_0 = elevation of PVC, ft (m)

E_t = elevation of PVT, ft (m)

x = distance of any point on the curve from the PVC, ft (m)

E_x = elevation of point x distant from PVC, ft (m)

x_s = distance from PVC to lowest point on a sag curve or highest point on a summit curve, ft (m)

E_s = elevation of lowest point on a sag curve or highest point on a summit curve, ft (m)

Equations of Parabolic Curves

In the parabolic-curve equations given next, algebraic quantities should always be used. Upward grades are positive and downward grades are negative.

$$R = \frac{G_2 - G_1}{L}$$

$$E_0 = V - \tfrac{1}{2} LG_1$$

$$E_x = E_0 + G_1 x + \tfrac{1}{2} Rx^2$$

$$x_s = -\frac{G_1}{R}$$

$$E_s = E_0 - \frac{G_1^2}{2R}$$

Note. If x_s is negative or if $x_s > L$, the curve does not have a high point or a low point.

HIGHWAY CURVES AND DRIVER SAFETY

For the safety and comfort of drivers, provision usually is made for gradual change from a tangent to the start of a circular curve.

As indicated in Fig. 11.4, typically the outer edge is raised first until the outer half of the cross section is level with the crown (point *B*). Then, the outer edge is raised farther until the cross section is straight (point *C*). From there on, the entire cross section is rotated until the full superelevation is attained (point *E*).

Superelevated roadway cross sections are typically employed on curves of rural highways and urban freeways. Superelevation is rarely used on local streets in residential, commercial, or industrial areas.

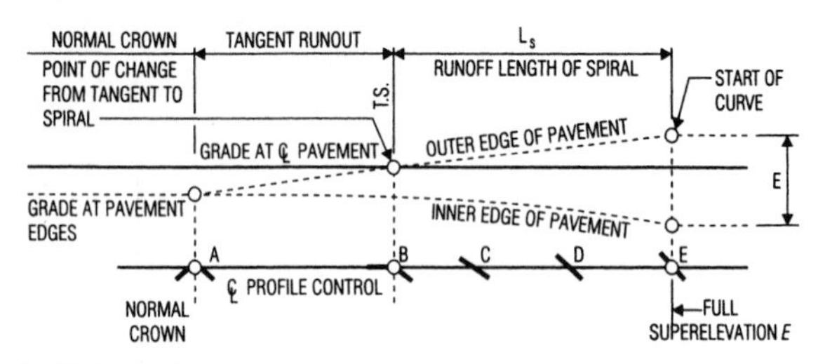

FIGURE 11.4 Superelevation variations along a spiral transition curve.

HIGHWAY ALIGNMENTS

Geometric design of a highway is concerned with horizontal and vertical alignment as well as the cross-sectional elements.

Horizontal alignment of a highway defines its location and orientation in plan view. Vertical alignment of a highway deals with its shape in profile. For a roadway with contiguous travel lanes, alignment can be conveniently represented by the centerline of the roadway.

Stationing

Distance along a horizontal alignment is measured in terms of stations. A full station is defined as 100 ft (30.48 m) and a half station as 50 ft (15.24 m). Station 100 + 50 is 150 ft (45.7 m) from the start of the alignment, station 0 + 00. A point 1492.27 ft (454.84 m) from 0 + 00 is denoted as 14 + 92.27, indicating a location 14 stations, 1400 ft

(426.72 m) plus 92.27 ft (28.12 m), from the starting point of the alignment. This distance is measured horizontally along the centerline of the roadway, whether it is a tangent, a curve, or a combination of these.

Stopping Sight Distance

This is the length of roadway needed between a vehicle and an arbitrary object (at some point down the road) to permit a driver to stop a vehicle safely before reaching the obstruction. This is not to be confused with passing sight distance, which *American Association of State Highway and Transportation Officials* (AASHTO) defines as the "length of roadway ahead visible to the driver." Figure 11.5 shows the parameters governing stopping sight distance on a crest vertical curve.

For crest vertical curves, AASHTO defines the minimum length L_{min}, ft (m), of crest vertical curves based on a required sight distance S, ft (m), as that given by

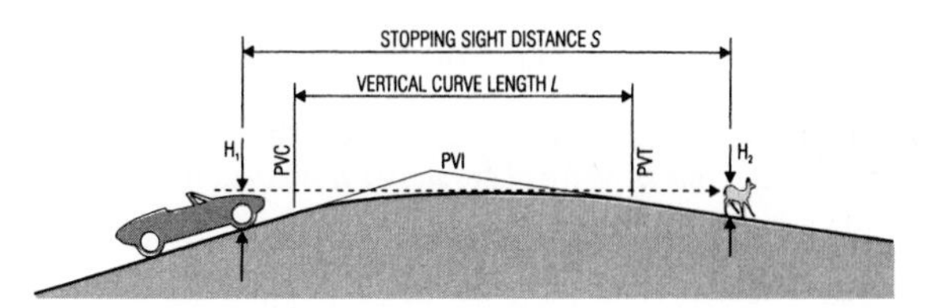

FIGURE 11.5 Stopping sight distance on a crest vertical curve.

$$L_{\min} = \frac{AS^2}{100\left(\sqrt{2H_1} + \sqrt{2H_2}\right)^2} \quad S < L$$

When eye height is 3.5 ft (1.07 m) and object height is 0.5 ft (0.152 m):

$$L_{\min} = \frac{AS^2}{1329} \quad S < L$$

Also, for crest vertical curves:

$$L_{\min} = 25 - \frac{200\left(\sqrt{H_1} + \sqrt{H_2}\right)^2}{AS^2} \quad S > L$$

When eye height is 3.5 ft (1.07 m) and object height 0.5 ft (0.152 m):

$$L_{\min} = 25 - \frac{1329}{AS^2} \quad S > L$$

where A = algebraic difference in grades, percent, of the tangents to the vertical curve

H_1 = eye height, ft (m), above the pavement

H_2 = object height, ft (m), above the pavement

Design controls for vertical curves can be established in terms of the rate of vertical curvature K defined by

$$K = \frac{L}{A}$$

where L = length, ft (m), of vertical curve and A is defined earlier. K is useful in determining the minimum sight distance, the length of a vertical curve from the PVC to the turning point (maximum point on a crest and minimum on a sag). This distance is found by multiplying K by the approach gradient.

Recommended values of K for various design velocities and stopping sight distances for crest and sag vertical curves are published by AASHTO.

STRUCTURAL NUMBERS FOR FLEXIBLE PAVEMENTS

The design of a flexible pavement or surface treatment expected to carry more than 50,000 repetitions of equivalent single 18-kip axle load (SAI) requires identification of a structural number SN that is used as a measure of the ability of the pavement to withstand anticipated axle loads. In the AASHTO design method, the structural number is defined by

$$SN = SN_1 + SN_2 + SN_3$$

where SN_1 = structural number for the surface course = $a_1 D_1$

a_1 = layer coefficient for the surface course

D_1 = actual thickness of the surface course, in (mm)

SN_2 = structural number for the base course = $a_2 D_2 m_2$

a_2 = layer coefficient for the base course

D_2 = actual thickness of the base course, in (mm)

m_2 = drainage coefficient for the base course

SN_3 = structural number for the subbase course $= a_3 D_3 m_3$

a_3 = layer coefficient for the subbase course

D_3 = actual thickness of the subbase course, in (mm)

m_3 = drainage coefficient for the subbase

The layer coefficients a_n are assigned to materials used in each layer to convert structural numbers to actual thickness. They are a measure of the relative ability of the materials to function as a structural component of the pavement. Many transportation agencies have their own values for these coefficients. As a guide, the layer coefficients may be 0.44 for asphaltic-concrete surface course, 0.14 for

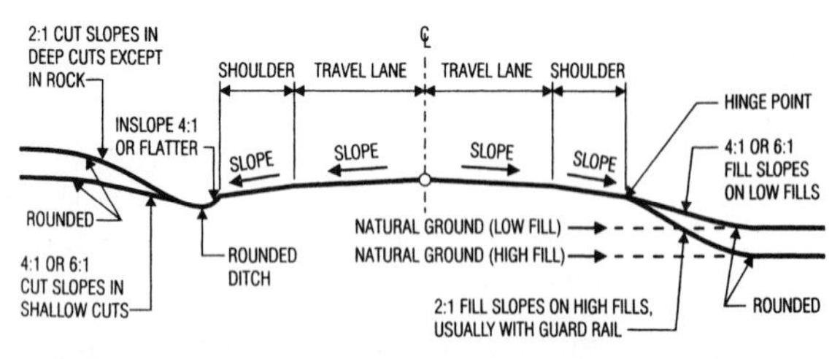

FIGURE 11.6 Typical two-lane highway with linear cross slopes.

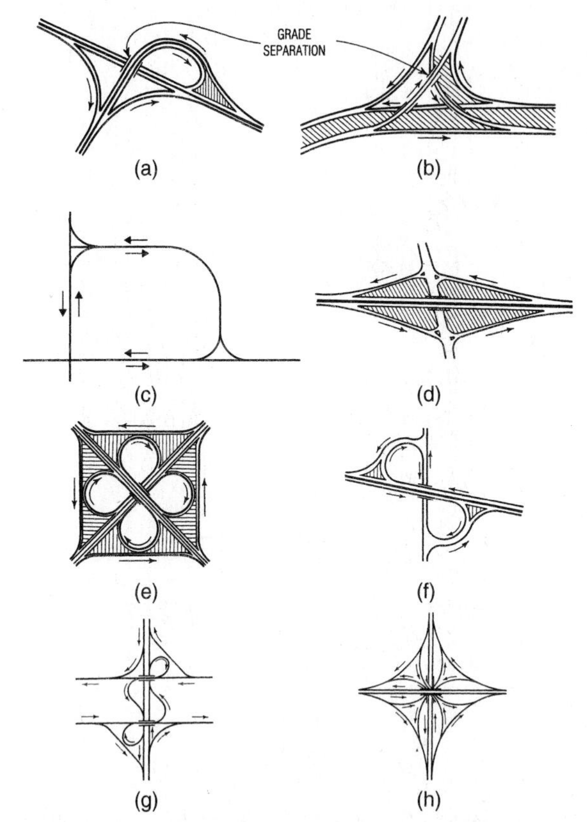

FIGURE 11.7 Types of interchanges for intersecting grade-separated highways. (*a*) T or trumpet; (*b*) Y or delta; (*c*) one quadrant; (*d*) diamond; (*e*) full cloverleaf; (*f*) partial cloverleaf; (*g*) semidirect; (*h*) all-directional four leg.

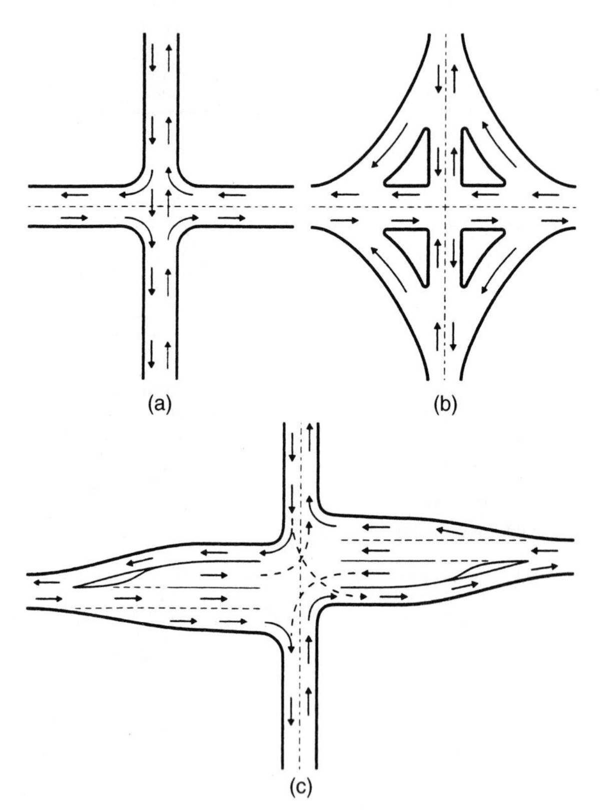

FIGURE 11.8 Highway turning lanes. (*a*) Unchannelized; (*b*) channelized; (*c*) flared.

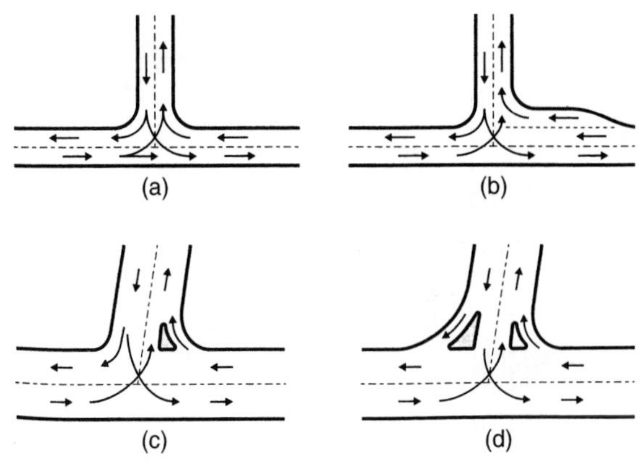

FIGURE 11.9 Highway turning lanes. (*a*) Unchannelized; (*b*) intersection with a right-turn lane; (*c*) intersection with a single-turning roadway; (*d*) channelized intersection with a pair of turning roadways.

crushed-stone base course, and 0.11 for sandy-gravel subbase course.

The thicknesses D_1, D_2, and D_3 should be rounded to the nearest $^1/_2$ in (12.7 mm). Selection of layer thicknesses usually is based on agency standards, maintainability of the pavement, and economic feasibility.

Figure 11.6 shows the linear cross slopes for a typical two-lane highway. Figure 11.7 shows the use of circular curves in a variety intersecting grade-separated highways.

Figure 11.8 shows the use of curves in at-grade four-leg intersections of highways. Figure 11.9 shows the use of curves in at-grade T (three-leg) intersections. Figure 11.10 shows street space and maneuvering space used for various parking positions.

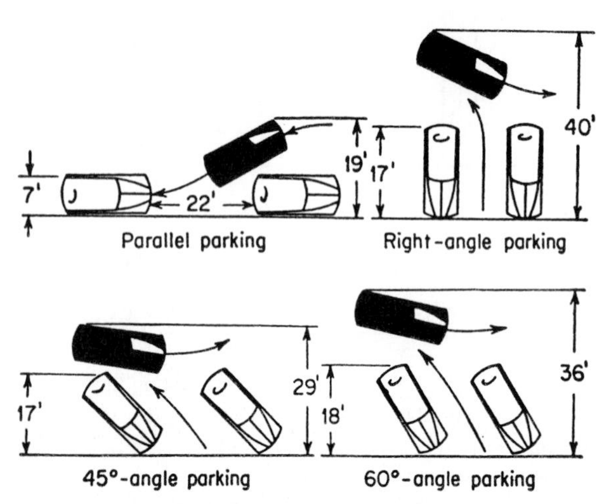

FIGURE 11.10 Street space and maneuvering space used for various parking positions. USCS (SI) equivalent units in ft (m): 7 (2.13), 17 (5.18), 18 (5.49), 19 (5.79), 22 (6.7), 29 (8.84), 36 (10.97), 40 (12.19).

TRANSITION (SPIRAL) CURVES

On starting around a horizontal circular curve, a vehicle and its contents are immediately subjected to centrifugal forces. The faster the vehicle enters the circle and the sharper the curvature is, the greater the influence on vehicles and drivers of the change from tangent to curve. When transition curves are not provided, drivers tend to create their own transition curves by moving laterally within their travel lane and sometimes the adjoining lane, a hazardous maneuver.

The minimum length L, ft (m), of a spiral may be computed from

$$L = \frac{3.15V^3}{RC}$$

where V = vehicle velocity, mi/h (km/h)

R = radius, ft (m), of the circular curve to which the spiral is joined

C = rate of increase of radial acceleration

An empirical value indicative of the comfort and safety involved, C values often used for highways range from 1 to 3. (For railroads, C is often taken as unity 1.) Another, more practical, method for calculating the minimum length of spiral required for use with circular curves is to base it on the required length for superelevation runoff.

DESIGNING HIGHWAY CULVERTS

A highway culvert is a pipelike drainage facility that allows water to flow under the road without impeding traffic. Corrugated and spiral steel pipe are popular for culverts because they can be installed quickly, have long life, are low in cost, and require little maintenance. With corrugated steel pipe, the *seam strength* must be adequate to withstand the ring-compression thrust from the total load supported by the pipe. This thrust C, lb/ft (N/m), of structure is

$$C = (LL + DL)\frac{S}{2}$$

where LL = live-load pressure, lb/ft^2 (N/m^2)

DL = dead-load pressure, lb/ft^2 (N/m^2)

S = span (or diameter), ft (m)

Handling and installation strength must be adequate to withstand shipping and placing of the pipe in the desired position at the highway job site. The handling strength is measured by a flexibility factor determined from

$$FF = \frac{D^2}{EI}$$

where D = pipe diameter or maximum span, in (mm)

E = modulus of elasticity of the pipe material, lb/in^2 (MPa)

I = moment of inertia per unit length of cross section of the pipe wall, in^4/in (mm^4/mm)

The ring-compression stress at which buckling becomes critical in the interaction zone for diameters less then $126.5r/K$ is

$$f_c = 45{,}000 - 1.406\left(\frac{KD}{r}\right)^2$$

For diameters greater than $126.5r/K$:

$$f_c = \frac{12E}{(KD/r)^2}$$

where f_c = buckling stress, lb/in^2 (MPa)

K = soil stiffness factor

D = pipe diameter or span, in (mm)

r = radius of gyration of pipe wall, in^4/in (mm^4/mm)

E = modulus of elasticity of pipe material, lb/in^2 (MPa)

Note. For excellent sidefill, compacted 90 to 95 percent of standard density, $K = 0.22$; for good sidefill, compacted to 85 percent of standard density, $K = 0.44$.

Conduit deflection is given by the Iowa formula. This formula gives the relative influence on the deflection of the pipe strength and the passive side pressure resisting horizontal movement of the pipe wall, or

$$\Delta_x = \frac{D_1 K W_c r^3}{EI + 0.061 E' r^3}$$

where Δ_x = horizontal deflection of pipe, in (mm)

D_1 = deflection lag factor

K = bedding constant (dependent on bedding angle)

W_c = vertical load per unit length of pipe, lb per linear in (N/mm)

r = mean radius of pipe, in (mm)

E = modulus of elasticity of pipe material, lb/in^2 (MPa)

I = moment of inertia per unit length of cross section of pipe wall, in^4/in (mm^4/mm)

E' = modulus of passive resistance of enveloping soil, lb/in^2 (MPa)

Soil modulus E' has not been correlated with the types of backfill and compaction. This limits the usefulness of the formula to analysis of installed structures that are under observation.

AMERICAN IRON AND STEEL INSTITUTE (AISI) DESIGN PROCEDURE

The design procedure for corrugated steel structures recommended in their *Handbook of Steel Drainage and Highway Construction Projects* is given below.

Backfill Density

Select a percentage of compaction of pipe backfill for design. The value chosen should reflect the importance and size of the structure and the quality that can reasonably be expected. The recommended value for routine use is 85 percent. This value usually applies to ordinary installations for which most specifications call for compaction to 90 percent. However, for more important structures in higher fill situations, consideration must be given to selecting higher quality backfill and requiring this quality for construction.

Design Pressure

When the height of cover is equal to, or greater than, the span or diameter of the structure, enter the load-factor chart (Fig. 11.11) to determine the percentage of the total load acting on the steel. For routine use, the 85 percent soil compaction provides a load factor $K = 0.86$. The total load is

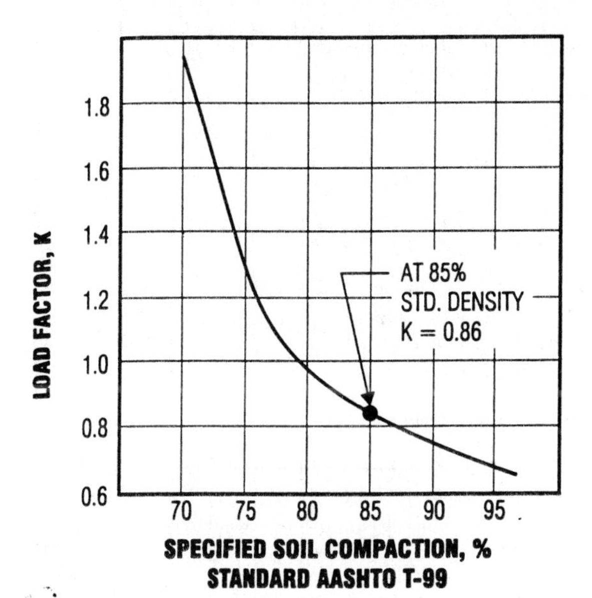

FIGURE 11.11 Load factors for corrugated steel pipe are plotted as a function of specified compaction of backfill.

multiplied by K to obtain the design pressure P_v acting on the steel. If the height of cover is less than one pipe diameter, the total load TL is assumed to act on the pipe, and $TL = P_v$; that is,

$$P_v = DL + LL + I \qquad H < S$$

When the height of cover is equal to, or greater than, one pipe diameter,

$$P_v = K(DL + LL + I) \qquad H \geq S$$

where P_v = design pressure, kip/ft^2 (MPa/m^2)

K = load factor

DL = dead load, kip/ft^2 (MPa/m^2)

LL = live load, kip/ft^2 (MPa/m^2)

I = impact, kip/ft^2 (MPa/m^2)

H = height of cover, ft (m)

S = span or pipe diameter, ft (m)

Ring Compression

The compressive thrust C, kip/ft (MPa/m), on the conduit wall equals the radial pressure P_v, kip/ft^2 (MPa/m^2), acting on the wall multiplied by the wall radius R, ft (m); or $C = P_vR$. This thrust, called ring compression, is the force carried by the steel. The ring compression is an axial load acting tangentially to the conduit wall (Fig. 11.12). For conventional structures in which the top arc approaches a semicircle, it is convenient to substitute half the span for the wall radius. Then,

$$C = P_v \frac{S}{2}$$

Allowable Wall Stress

The ultimate compression in the pipe wall is expressed by Eqs. (11.1) and (11.2) that follow. The ultimate wall stress is equal to the specified minimum yield point of the steel

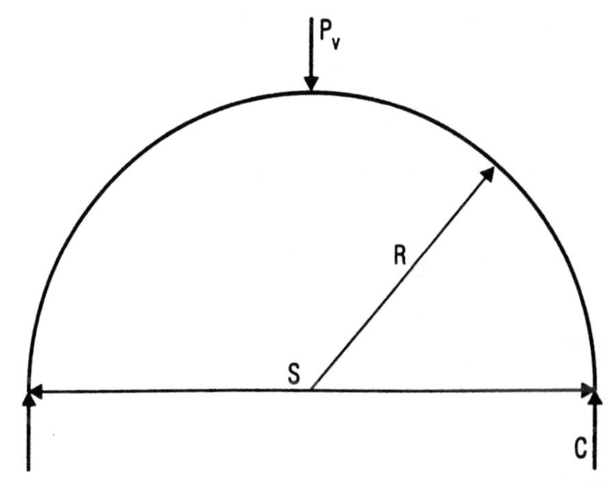

FIGURE 11.12 Radial pressure, P_v, on the wall of a curved conduit is resisted by compressive thrust C.

and applies to the zone of wall crushing or yielding. Equation (11.1) applies to the interaction zone of yielding and ring buckling; Eq. (11.2) applies to the ring-buckling zone.

When the ratio D/r of pipe diameter—or span D, in (mm), to radius of gyration r, in (mm), of the pipe cross section—does not exceed 294, the ultimate wall stress may be taken as equal to the steel yield strength:

$$F_b = F_y = 33 \text{ ksi } (227.4 \text{ MPa})$$

When D/r exceeds 294 but not 500, the ultimate wall stress, ksi (MPa), is given by

$$F_b = 40 - 0.000081 \left(\frac{D}{r}\right)^2 \qquad (11.1)$$

When D/r is more than 500,

$$F_b = \frac{4.93 \times 10^6}{(D/r)^2} \qquad (11.2)$$

A safety factor of 2 is applied to the ultimate wall stress to obtain the design stress F_c, ksi (MPa):

$$F_c = \frac{F_b}{2} \qquad (11.3)$$

Wall Thickness

Required wall area A, in^2/ft (mm^2/m), of width, is computed from the calculated compression C in the pipe wall and the allowable stress F_c:

$$A = \frac{C}{F_c} \qquad (11.4)$$

From the AISI table for underground conduits, select the wall thickness that provides the required area with the same corrugation used for selection of the allowable stress.

Check Handling Stiffness

Minimum pipe stiffness requirements for practical handling and installation, without undue care or bracing, have been established through experience. The resulting flexibility factor FF limits the size of each combination of corrugation pitch and metal thickness:

$$FF = \frac{D^2}{EI} \qquad (11.5)$$

where E = modulus of elasticity, ksi (MPa), of steel = 30,000 ksi (206,850 MPa); and I = moment of inertia of wall, in^4/in (mm^4/mm).

The following maximum values of FF are recommended for ordinary installations:

FF = 0.0433 for factory-made pipe with less than a 120-in (30.48-cm) diameter and with riveted, welded, or helical seams

FF = 0.0200 for field-assembled pipe with over a 120-in (30.48-cm) diameter or with bolted seams

Higher values can be used with special care or where experience indicates. Trench condition, as in sewer design, can be one such case; use of aluminum pipe is another. For example, the flexibility factor permitted for aluminum pipe in some national specifications is more than twice that recommended here for steel because aluminum has only one-third the stiffness of steel, the modulus for aluminum being about 10,000 vs. 30,000 ksi (68,950 vs. 206,850 MPa) for steel. Where a high degree of flexibility is acceptable for aluminum, it is equally acceptable for steel.

Check Bolted Seams

Standard factory-made pipe seams are satisfactory for all designs within the maximum allowable wall stress of 16.5 ksi (113.8 MPa). Seams bolted in the shop or field, however, continue to be evaluated on the basis of test values for uncurved, unsupported columns. A bolted seam (standard for structural plate) must have a test strength of twice the design load in the pipe wall.

CHAPTER 12

HYDRAULICS AND WATERWORKS FORMULAS

To simplify using the formulas in this chapter, Table 12.1 presents symbols, nomenclature, and *United States Customary System* (USCS) and *System International* (SI) units found in each expression.

CAPILLARY ACTION

Capillarity is due to both the cohesive forces between liquid molecules and adhesive forces of liquid molecules. It shows up as the difference in liquid surface elevations between the inside and outside of a small tube that has one end submerged in the liquid (Fig. 12.1).

Capillarity is commonly expressed as the height of this rise. In equation form,

$$h = \frac{2\sigma \cos \theta}{(w_1 - w_2)r}$$

where h = capillary rise, ft (m)

σ = surface tension, lb/ft (N/m)

w_1 and w_2 = specific weights of fluids below and above meniscus, respectively, lb/ft (N/m)

θ = angle of contact

r = radius of capillary tube, ft (m)

Capillarity, like surface tension, decreases with increasing temperature. Its temperature variation, however, is small and insignificant in most problems.

TABLE 12.1 Symbols, Terminology, Dimensions, and Units Used in Water Engineering

Symbol	Terminology	Dimensions	USCS units	SI units
A	Area	L^2	ft^2	mm^2
C	Chezy roughness coefficient	$L^{1/2}/T$	ft^5/s	$m^{0.5}/s$
C_1	Hazen–Williams roughness coefficient	$L^{0.37}/T$	$ft^{0.37}/s$	$m^{0.37}/s$
d	Depth	L	ft	m
d_c	Critical depth	L	ft	m
D	Diameter	L	ft	m
E	Modulus of elasticity	F/L^2	lb/in^2	MPa
F	Force	F	lb	N
g	Acceleration due to gravity	L/T^2	ft/s^2	m/s^2
H	Total head, head on weir	L	ft	m
h	Head or height	L	ft	m
h_f	Head loss due to friction	L	ft	m
L	Length	L	ft	m
M	Mass	FT^2/L	$lb \cdot s^2/ft$	Ns^2/m
n	Manning's roughness coefficient	$T/L^{1/3}$	$s/ft^{1/3}$	$s/m^{1/3}$
P	Perimeter, weir height	L	ft	m
P	Force due to pressure	F	lb	N
p	Pressure	F/L^2	psf	MPa

TABLE 12.1 Symbols, Terminology, Dimensions, and Units Used in Water Engineering *(Continued)*

Symbol	Terminology	Dimensions	USCS units	SI units
Q	Flow rate	L^3/T	ft^3/s	m^3/s
q	Unit flow rate	$L^3/T \cdot L$	$ft^3/(s \cdot ft)$	$m^3/s \cdot m$
r	Radius	L	ft	m
R	Hydraulic radius	L	ft	m
T	Time	T	s	s
t	Time, thickness	T, L	s, ft	s, m
V	Velocity	L/T	ft/s	m/s
W	Weight	F	lb	kg
w	Specific weight	F/L^3	lb/ft^3	kg/m^3
y	Depth in open channel, distance from solid boundary	L	ft	m
Z	Height above datum	L	ft	m
ϵ	Size of roughness	L	ft	m
μ	Viscosity	FT/L^2	$lb \cdot s/ft$	$kg \cdot s/m$
ν	Kinematic viscosity	L^2/T	ft^2/s	m^2/s
ρ	Density	FT^2/L^4	$lb \cdot s^2/ft^4$	$kg \cdot s^2/m^4$
σ	Surface tension	F/L	lb/ft	kg/m
τ	Shear stress	F/L^2	lb/in^2	MPa

Symbols for dimensionless quantities

Symbol	Quantity
C	Weir coefficient, coefficient of discharge
C_c	Coefficient of contraction
C_v	Coefficient of velocity
$\mathbf{F}$	Froude number
f	Darcy–Weisbach friction factor
K	Head-loss coefficient
$\mathbf{R}$	Reynolds number
S	Friction slope—slope of energy grade line
S_c	Critical slope
η	Efficiency
Sp. gr.	Specific gravity

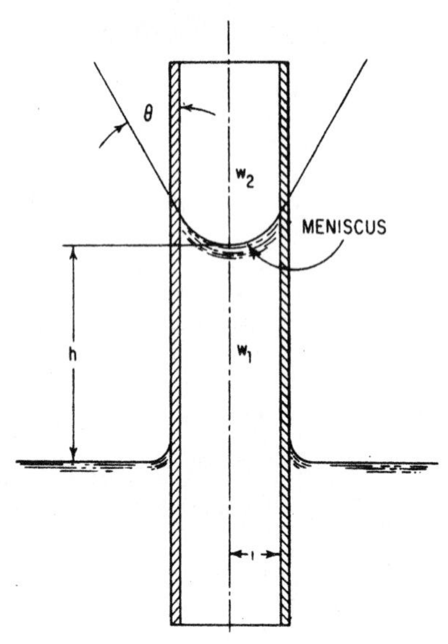

FIGURE 12.1 Capillary action raises water in a small-diameter tube. Meniscus, or liquid surface, is concave upward.

VISCOSITY

Viscosity μ of a fluid, also called the *coefficient of viscosity*, *absolute viscosity*, or *dynamic viscosity*, is a measure of its resistance to flow. It is expressed as the ratio of the tangential shearing stresses between flow layers to the rate of change of velocity with depth:

$$\mu = \frac{\tau}{dV/dy}$$

where τ = shearing stress, lb/ft^2 (N/m^2)

V = velocity, ft/s (m/s)

y = depth, ft (m)

Viscosity decreases as temperature increases but may be assumed independent of changes in pressure for the majority of engineering problems. Water at 70°F (21.1°C) has a viscosity of 0.00002050 lb·s/ft^2 (0.00098 N·s/m^2).

Kinematic viscosity ν is defined as viscosity μ divided by density ρ. It is so named because its units, ft^2/s (m^2/s), are a combination of the kinematic units of length and time. Water at 70°F (21.1°C) has a kinematic viscosity of 0.00001059 ft^2/s (0.000001 Nm^2/s).

In hydraulics, viscosity is most frequently encountered in the calculation of Reynolds number to determine whether laminar, transitional, or completely turbulent flow exists.

PRESSURE ON SUBMERGED CURVED SURFACES

The hydrostatic pressure on a submerged curved surface (Fig. 12.2) is given by

$$P = \sqrt{P_H^2 + P_V^2}$$

where P = total pressure force on the surface

P_H = force due to pressure horizontally

P_V = force due to pressure vertically

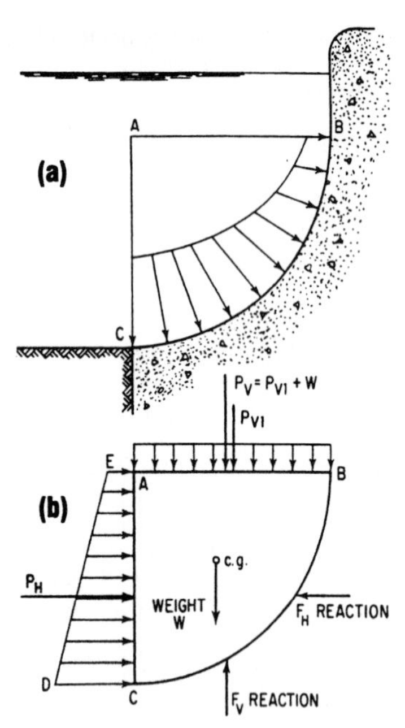

FIGURE 12.2 Hydrostatic pressure on a submerged curved surface. (*a*) Pressure variation over the surface. (*b*) Free-body diagram.

FUNDAMENTALS OF FLUID FLOW

For fluid energy, the law of conservation of energy is represented by the *Bernoulli equation*:

$$Z_1 + \frac{p_1}{w} + \frac{V_1^2}{2g} = Z_2 + \frac{p_2}{w} + \frac{V_2^2}{2g}$$

where Z_1 = elevation, ft (m), at any point 1 of flowing fluid above an arbitrary datum

Z_2 = elevation, ft (m), at downstream point in fluid above same datum

p_1 = pressure at 1, lb/ft^2 (kPa)

p_2 = pressure at 2, lb/ft^2 (kPa)

w = specific weight of fluid, lb/ft^3 (kg/m^3)

V_1 = velocity of fluid at 1, ft/s (m/s)

V_2 = velocity of fluid at 2, ft/s (m/s)

g = acceleration due to gravity, 32.2 ft/s^2 (9.81 m/s^2)

The left side of the equation sums the total energy per unit weight of fluid at 1, and the right side, the total energy per unit weight at 2. The preceding equation applies only to an ideal fluid. Its practical use requires a term to account for the decrease in total head, ft (m), through friction. This term h_f, when added to the downstream side, yields the form of the Bernoulli equation most frequently used:

$$Z_1 + \frac{p_1}{w} + \frac{V_1^2}{2g} = Z_2 + \frac{p_2}{w} + \frac{V_2^2}{2g} + h_f$$

The energy contained in an elemental volume of fluid thus is a function of its elevation, velocity, and pressure (Fig. 12.3). The energy due to elevation is the potential energy and equals WZ_a, where W is the weight, lb (kg), of the fluid in the elemental volume and Z_a is its elevation, ft (m), above some arbitrary datum. The energy due to velocity is the kinetic energy. It equals $WV_a^2/2g$, where V_a is the

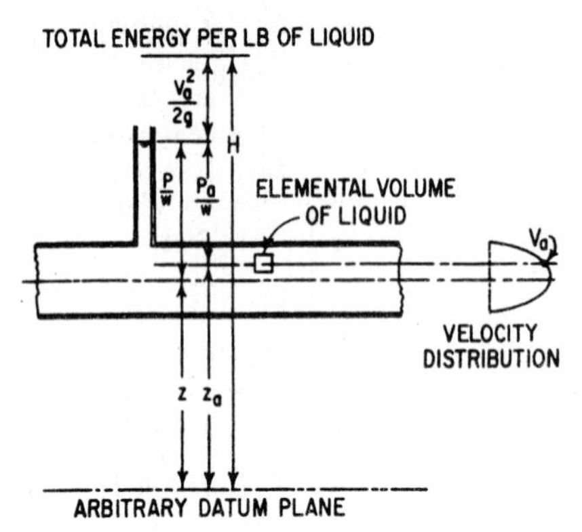

FIGURE 12.3 Energy in a liquid depends on elevation, velocity, and pressure.

velocity, ft/s (m/s). The pressure energy equals Wp_a/w, where p_a is the pressure, lb/ft^2 (kg/kPa), and w is the specific weight of the fluid, lb/ft^3 (kg/m^3). The total energy in the elemental volume of fluid is

$$E = WZ_a + \frac{Wp_a}{w} + \frac{WV_a^2}{2g}$$

Dividing both sides of the equation by W yields the energy per unit weight of flowing fluid, or the *total head* ft (m):

$$H = Z_a + \frac{p_a}{w} + \frac{V_a^2}{2g}$$

p_a/w is called *pressure head*; $V_a^2/2g$, *velocity head.*

As indicated in Fig. 12.3, $Z + p/w$ is constant for any point in a cross section and normal to the flow through a pipe or channel. Kinetic energy at the section, however, varies with velocity. Usually, $Z + p/w$ at the midpoint and the average velocity at a section are assumed when the Bernoulli equation is applied to flow across the section or when total head is to be determined. *Average velocity*, ft/s (m/s) $= Q/A$, where Q is the quantity of flow, ft^3/s (m^3/s), across the area of the section A, ft^2 (m^2).

Momentum is a fundamental concept that must be considered in the design of essentially all waterworks facilities involving flow. A change in momentum, which may result from a change in velocity, direction, or magnitude of flow, is equal to the impulse, the force F acting on the fluid times the period of time dt over which it acts (Fig. 12.4). Dividing the total change in momentum by the time interval over which the change occurs gives the momentum equation, or impulse-momentum equation:

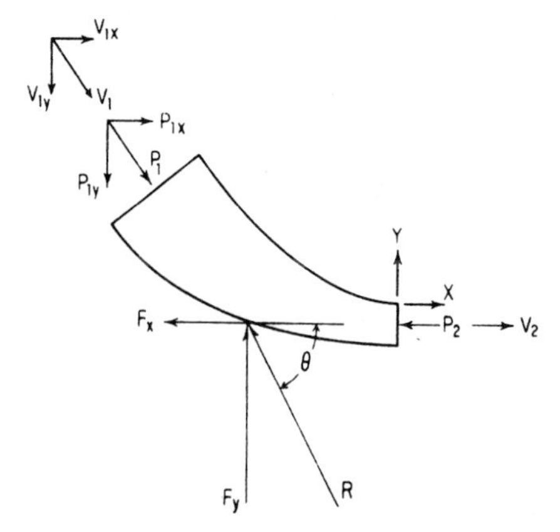

FIGURE 12.4 Force diagram for momentum.

$$F_x = pQ\,\Delta V_x$$

where F_x = summation of all forces in X direction per unit time causing change in momentum in X direction, lb (N)

ρ = density of flowing fluid, lb·s²/ft⁴ (kg·s²/m⁴) (specific weight divided by g)

Q = flow rate, ft³/s (m³/s)

ΔV_x = change in velocity in X direction, ft/s (m/s)

Similar equations may be written for the Y and Z directions. The impulse–momentum equation often is used in conjunction with the Bernoulli equation but may be used separately.

SIMILITUDE FOR PHYSICAL MODELS

A physical model is a system whose operation can be used to predict the characteristics of a similar system, or prototype, usually more complex or built to a much larger scale.

Ratios of the forces of gravity, viscosity, and surface tension to the force of inertia are designated, Froude number, Reynolds number, and Weber number, respectively. Equating the Froude number of the model and the Froude number of the prototype ensures that the gravitational and inertial forces are in the same proportion. Similarly, equating the Reynolds numbers of the model and prototype ensures that the viscous and inertial forces are in the same proportion. Equating the Weber numbers ensures proportionality of surface tension and inertial forces.

The *Froude number* is

$$\mathbf{F} = \frac{V}{\sqrt{Lg}}$$

where $\mathbf{F}$ = Froude number (dimensionless)

V = velocity of fluid, ft/s (m/s)

L = linear dimension (characteristic, such as depth or diameter), ft (m)

g = acceleration due to gravity, 32.2 ft/s^2 (9.81 m/s^2)

For hydraulic structures, such as spillways and weirs, where there is a rapidly changing water-surface profile, the two predominant forces are inertia and gravity. Therefore, the Froude numbers of the model and prototype are equated:

$$\mathbf{F}_{\mathrm{m}} = \mathbf{F}_{\mathrm{p}} \qquad \frac{V_m}{\sqrt{L_m g}} = \frac{V_p}{\sqrt{L_p g}}$$

where subscript m applies to the model and p to the prototype.

The *Reynolds number* is

$$\mathbf{R} = \frac{VL}{\nu}$$

$\mathbf{R}$ is dimensionless, and ν is the kinematic viscosity of fluid, ft^2/s (m^2/s). The Reynolds numbers of model and prototype are equated when the viscous and inertial forces are predominant. Viscous forces are usually predominant when flow occurs in a closed system, such as pipe flow where there is no free surface. The following relations are obtained by equating Reynolds numbers of the model and prototype:

$$\frac{V_m L_m}{v_m} = \frac{V_p L_p}{v_p} \qquad V_r = \frac{v_r}{L_r}$$

The variable factors that fix the design of a true model when the Reynolds number governs are the length ratio and the viscosity ratio.

The *Weber number* is

$$\mathbf{W} = \frac{V^2 L\rho}{\sigma}$$

where ρ = density of fluid, lb·s²/ft⁴ (kg·s²/m⁴) (specific weight divided by g); and σ = surface tension of fluid, lb/ft² (kPa).

The Weber numbers of model and prototype are equated in certain types of wave studies.

For the flow of water in open channels and rivers where the friction slope is relatively flat, model designs are often based on the Manning equation. The relations between the model and prototype are determined as follows:

$$\frac{V_m}{V_p} = \frac{(1.486/n_m)R_m^{2/3}S_m^{1/2}}{(1.486/n_p)R_p^{2/3}S_p^{1/2}}$$

where n = Manning roughness coefficient ($T/L^{1/3}$, T representing time)

R = hydraulic radius (L)

S = loss of head due to friction per unit length of conduit (dimensionless)

= slope of energy gradient

For true models, $S_r = 1$, $R_r = L_r$. Hence,

$$V_r = \frac{L_r^{2/3}}{n_r}$$

In models of rivers and channels, it is necessary for the flow to be turbulent. The U.S. Waterways Experiment Station has determined that flow is turbulent if

$$\frac{VR}{\nu} \geq 4000$$

where V = mean velocity, ft/s (m/s)

R = hydraulic radius, ft (m)

ν = kinematic viscosity, ft^2/s (m^2/s)

If the model is to be a true model, it may have to be uneconomically large for the flow to be turbulent.

FLUID FLOW IN PIPES

Laminar Flow

In laminar flow, fluid particles move in parallel layers in one direction. The parabolic velocity distribution in laminar flow, shown in Fig. 12.5, creates a shearing stress $\tau = \mu\, dV/dy$, where dV/dy is the rate of change of velocity with depth, and μ is the coefficient of viscosity. As this shearing stress increases, the viscous forces become unable to damp out disturbances, and turbulent flow results. The region of change is dependent on the fluid velocity, density, and viscosity and the size of the conduit.

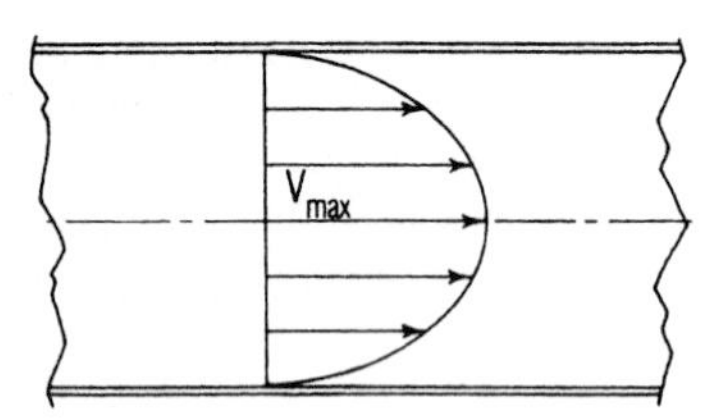

FIGURE 12.5 Velocity distribution for lamellar flow in a circular pipe is parabolic. Maximum velocity is twice the average velocity.

A dimensionless parameter called the Reynolds number has been found to be a reliable criterion for the determination of laminar or turbulent flow. It is the ratio of inertial forces/viscous forces, and is given by

$$\mathbf{R} = \frac{VD\rho}{\mu} = \frac{VD}{\nu}$$

where V = fluid velocity, ft/s (m/s)

D = pipe diameter, ft (m)

ρ = density of fluid, lb·s²/ft⁴ (kg·s²/m⁴) (specific weight divided by g, 32.2 ft/s²)

μ = viscosity of fluid lb·s/ft² (kg·s/m²)

$\nu = \mu/\rho$ = kinematic viscosity, ft²/s (m²/s)

For a Reynolds number less than 2000, flow is laminar in circular pipes. When the Reynolds number is greater than 2000, laminar flow is unstable; a disturbance is probably magnified, causing the flow to become turbulent.

In laminar flow, the following equation for head loss due to friction can be developed by considering the forces acting on a cylinder of fluid in a pipe:

$$h_f = \frac{32\mu LV}{D^2 \rho g} = \frac{32\mu LV}{D^2 w}$$

where h_f = head loss due to friction, ft (m)

L = length of pipe section considered, ft (m)

g = acceleration due to gravity, 32.2 ft/s^2 (9.81 m/s^2)

w = specific weight of fluid, lb/ft^3 (kg/m^3)

Substitution of the Reynolds number yields

$$h_f = \frac{64}{\mathbf{R}} \frac{L}{D} \frac{V^2}{2g}$$

For laminar flow, the preceding equation is identical to the Darcy–Weisbach formula because, in laminar flow, the friction $f = 64/\mathbf{R}$.

Turbulent Flow

In turbulent flow, the inertial forces are so great that viscous forces cannot dampen out disturbances caused primarily by the surface roughness. These disturbances create eddies, which have both a rotational and translational velocity. The translation of these eddies is a mixing action that affects an interchange of momentum across the cross section of the conduit. As a result, the velocity distribution is more uniform, as shown in Fig. 12.6. Experimentation in turbulent flow has shown that

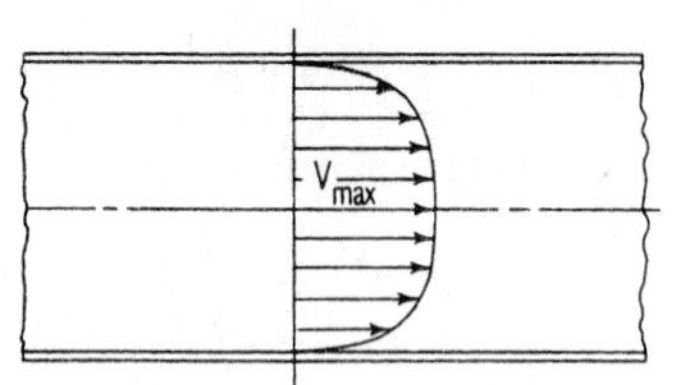

FIGURE 12.6 Velocity distribution for turbulent flow in a circular pipe is more nearly uniform than that for lamellar flow.

The head loss varies directly as the length of the pipe.

The head loss varies almost as the square of the velocity.

The head loss varies almost inversely as the diameter.

The head loss depends on the surface roughness of the pipe wall.

The head loss depends on the fluid density and viscosity.

The head loss is independent of the pressure.

Darcy–Weisbach Formula

One of the most widely used equations for pipe flow, the Darcy–Weisbach formula satisfies the condition described in the preceding section and is valid for laminar or turbulent flow in all fluids:

$$h_f = f \frac{L}{D} \frac{V^2}{2g}$$

where h_f = head loss due to friction, ft (m)

f = friction factor (see an engineering handbook)

L = length of pipe, ft (m)

D = diameter of pipe, ft (m)

V = velocity of fluid, ft/s (m/s)

g = acceleration due to gravity, 32.2 ft/s^2 (9.81 m/s^2)

It employs the Moody diagram for evaluating the friction factor f. (Moody, L. F., "Friction Factors for Pipe Flow," *Transactions of the American Society of Mechanical Engineers*, November 1944.)

Because the preceding equation is dimensionally homogeneous, it can be used with any consistent set of units without changing the value of the friction factor.

Roughness values ϵ, ft (m), for use with the Moody diagram to determine the Darcy–Weisbach friction factor f are listed in engineering handbooks.

The following formulas were derived for head loss in waterworks design and give good results for water-transmission and -distribution calculations. They contain a factor that depends on the surface roughness of the pipe material. The accuracy of these formulas is greatly affected by the selection of the roughness factor, which requires experience in its choice.

Chezy Formula

This equation holds for head loss in conduits and gives reasonably good results for high Reynolds numbers:

$$V = C\sqrt{RS}$$

where V = velocity, ft/s (m/s)

C = coefficient, dependent on surface roughness of conduit

S = slope of energy grade line or head loss due to friction, ft/ft (m/m) of conduit

R = hydraulic radius, ft (m)

Hydraulic radius of a conduit is the cross-sectional area of the fluid in it divided by the perimeter of the wetted section.

Manning's Formula

Through experimentation, Manning concluded that the C in the Chezy equation should vary as $R^{1/6}$:

$$C = \frac{1.486R^{1/6}}{n}$$

where n = coefficient, dependent on surface roughness. (Although based on surface roughness, n in practice is sometimes treated as a lumped parameter for all head losses.) Substitution gives

$$V = \frac{1.486}{n} R^{2/3}S^{1/2}$$

On substitution of $D/4$, where D is the pipe diameter, for the hydraulic radius of the pipe, the following equations are obtained for pipes flowing full:

$$V = \frac{0.590}{n} D^{2/3}S^{1/2}$$

$$Q = \frac{0.463}{n} D^{8/3}S^{1/2}$$

$$h_f = 4.66n^2 \frac{LQ^2}{D^{16/3}}$$

$$D = \left(\frac{2.159Qn}{S^{1/2}}\right)^{3/8}$$

where Q = flow, ft^3/s (m^3/s).

Hazen–Williams Formula

This is one of the most widely used formulas for pipe-flow computations of water utilities, although it was developed for both open channels and pipe flow:

$$V = 1.318C_1R^{0.63}S^{0.54}$$

For pipes flowing full:

$$V = 0.55C_1D^{0.63}S^{0.54}$$

$$Q = 0.432C_1D^{2.63}S^{0.54}$$

$$h_f = \frac{4.727}{D^{4.87}} L \left(\frac{Q}{C_1}\right)^{1.85}$$

$$D = \frac{1.376}{S^{0.205}} \left(\frac{Q}{C_1}\right)^{0.38}$$

where V = velocity, ft/s (m/s)

C_1 = coefficient, dependent on surface roughness (given in engineering handbooks)

R = hydraulic radius, ft (m)

S = head loss due to friction, ft/ft (m/m) of pipe

D = diameter of pipe, ft (m)

L = length of pipe, ft (m)

Q = discharge, ft^3/s (m^3/s)

h_f = friction loss, ft (m)

Figure 12.7 shows a typical three-reservoir problem. The elevations of the hydraulic grade lines for the three pipes are equal at point D. The Hazen–Williams equation for friction loss can be written for each pipe meeting at D. With the continuity equation for quantity of flow, there are as many equations as there are unknowns:

$$Z_a = Z_d + \frac{pD}{w} + \frac{4.727L_A}{D_A^{4.87}}\left(\frac{Q_A}{C_A}\right)^{1.85}$$

$$Z_b = Z_d + \frac{P_D}{w} + \frac{4.727L_B}{D_B^{4.87}}\left(\frac{Q_B}{C_B}\right)^{1.85}$$

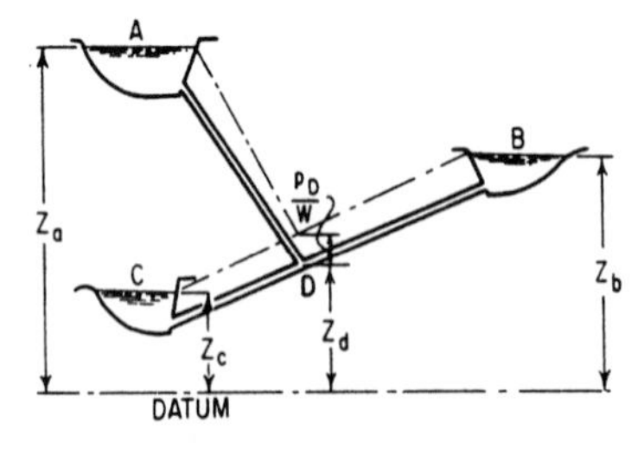

FIGURE 12.7 Flow between reservoirs.

$$Z_c = Z_d + \frac{P_D}{w} + \frac{4.727L_C}{D_C^{4.87}}\left(\frac{Q_C}{C_C}\right)^{1.85}$$

$$Q_A + Q_B = Q_C$$

where p_D = pressure at D, and w = unit weight of liquid.

PRESSURE (HEAD) CHANGES CAUSED BY PIPE SIZE CHANGE

Energy losses occur in pipe contractions, bends, enlargements, and valves and other pipe fittings. These losses can usually be neglected if the length of the pipeline is greater than 1500 times the pipe diameter. However, in short pipelines, because these losses may exceed the friction losses, minor losses must be considered.

Sudden Enlargements

The following equation for the head loss, ft (m), across a sudden enlargement of pipe diameter has been determined analytically and agrees well with experimental results:

$$h_L = \frac{(V_1 - V_2)^2}{2g}$$

where V_1 = velocity before enlargement, ft/s (m/s)

V_2 = velocity after enlargement, ft/s (m/s)

g = 32.2 ft/s^2 (9.81 m/s^2)

It was derived by applying the Bernoulli equation and the momentum equation across an enlargement.

Another equation for the head loss caused by sudden enlargements was determined experimentally by Archer. This equation gives slightly better agreement with experimental results than the preceding formula:

$$h_L = \frac{1.1(V_1 - V_2)^{1.92}}{2g}$$

A special application of these two preceding formulas is the discharge from a pipe into a reservoir. The water in the reservoir has no velocity, so a full velocity head is lost.

Gradual Enlargements

The equation for the head loss due to a gradual conical enlargement of a pipe takes the following form:

$$h_L = \frac{K(V_1 - V_2)^2}{2g}$$

where K = loss coefficient, as given in engineering handbooks.

Sudden Contraction

The following equation for the head loss across a sudden contraction of a pipe was determined by the same type of analytic studies as

$$h_L = \left(\frac{1}{C_c} - 1\right)^2 \frac{V^2}{2g}$$

where C_c = coefficient of contraction; and V = velocity in smaller diameter pipe, ft/s (m/s). This equation gives best results when the head loss is greater than 1 ft (0.3 m).

Another formula for determining the loss of head caused by a sudden contraction, determined experimentally by Brightmore, is

$$h_L = \frac{0.7(V_1 - V_2)^2}{2g}$$

This equation gives best results if the head loss is less than 1 ft (0.3 m).

A special case of sudden contraction is the entrance loss for pipes. Some typical values of the loss coefficient K in $h_L = KV^2/2g$, where V is the velocity in the pipe, are presented in engineering handbooks.

Bends and Standard Fitting Losses

The head loss that occurs in pipe fittings, such as valves and elbows, and at bends is given by

$$h_L = \frac{KV^2}{2g}$$

To obtain losses in bends other than 90°, the following formula may be used to adjust the K values:

$$K' = K\sqrt{\frac{\Delta}{90}}$$

where Δ = deflection angle, degrees. K values are given in engineering handbooks.

FLOW THROUGH ORIFICES

An orifice is an opening with a closed perimeter through which water flows. Orifices may have any shape, although they are usually round, square, or rectangular.

Orifice Discharge into Free Air

Discharge through a sharp-edged orifice may be calculated from

$$Q = Ca\sqrt{2gh}$$

where Q = discharge, ft^3/s (m^3/s)

C = coefficient of discharge

a = area of orifice, ft^2 (m^2)

g = acceleration due to gravity, ft/s^2 (m/s^2)

h = head on horizontal center line of orifice, ft (m)

Coefficients of discharge C are given in engineering handbooks for low velocity of approach. If this velocity is significant, its effect should be taken into account. The preceding formula is applicable for any head for which the coefficient of discharge is known. For low heads, measuring the head from the center line of the orifice is not theoretically correct; however, this error is corrected by the C values.

The *coefficient of discharge* C is the product of the coefficient of velocity C_v and the coefficient of contraction C_c. The *coefficient of velocity* is the ratio obtained by dividing the actual velocity at the *vena contracta* (contraction of the jet discharged) by the theoretical velocity. The theoretical

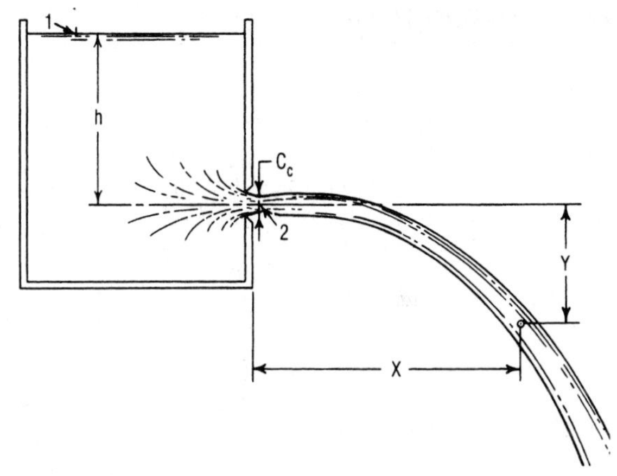

FIGURE 12.8 Fluid jet takes a parabolic path.

velocity may be calculated by writing Bernoulli's equation for points 1 and 2 in Fig. 12.8.

$$\frac{V_1^2}{2g} + \frac{p_1}{w} + Z_1 = \frac{V_2^2}{2g} + \frac{p_2}{w} + Z_2$$

With the reference plane through point 2, $Z_1 = h$, $V_1 = 0$, $p_1/w = p_2/w = 0$, and $Z_2 = 0$, the preceding formula becomes

$$V_2 = \sqrt{2gh}$$

The *coefficient of contraction* C_c is the ratio of the smallest area of the jet, the vena contracta, to the area of the orifice. Contraction of a fluid jet occurs if the orifice is square edged and so located that some of the fluid approaches the orifice at an angle to the direction of flow through the orifice.

Submerged Orifices

Flow through a submerged orifice may be computed by applying Bernoulli's equation to points 1 and 2 in Fig. 12.9:

$$V_2 = \sqrt{2g\left(h_1 - h_2 + \frac{V_1^2}{2g} - h_L\right)}$$

where h_L = losses in head, ft (m), between 1 and 2.

By assuming $V_1 \approx 0$, setting $h_1 - h_2 = \Delta h$, and using a coefficient of discharge C to account for losses, the following formula is obtained:

$$Q = Ca\sqrt{2g\,\Delta h}$$

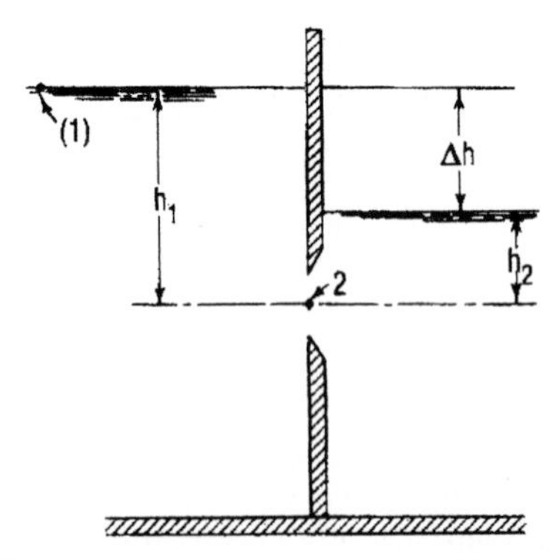

FIGURE 12.9 Discharge through a submerged orifice.

Values of C for submerged orifices do not differ greatly from those for nonsubmerged orifices.

Discharge under Falling Head

The flow from a reservoir or vessel when the inflow is less than the outflow represents a condition of falling head. The time required for a certain quantity of water to flow from a reservoir can be calculated by equating the volume of water that flows through the orifice or pipe in time dt to the volume decrease in the reservoir. If the area of the reservoir is constant,

$$t = \frac{2A}{Ca\sqrt{2g}}\left(\sqrt{h_1} - \sqrt{h_2}\right)$$

where h_1 = head at the start, ft (m)

h_2 = head at the end, ft (m)

t = time interval for head to fall from h_1 to h_2, s

FLUID JETS

Where the effect of air resistance is small, a fluid discharged through an orifice into the air follows the path of a projectile. The initial velocity of the jet is

$$V_0 = C_v\sqrt{2gh}$$

where h = head on center line of orifice, ft (m), and C_v = coefficient of velocity.

The direction of the initial velocity depends on the orientation of the surface in which the orifice is located. For simplicity, the following equations were determined assuming the orifice is located in a vertical surface (see Fig. 12.8). The velocity of the jet in the X direction (horizontal) remains constant:

$$V_x = V_0 = C_v \sqrt{2gh}$$

The velocity in the Y direction is initially zero and thereafter a function of time and the acceleration of gravity:

$$V_y = gt$$

The X coordinate at time t is

$$X = V_x t = tC_v \sqrt{2gh}$$

The Y coordinate is

$$Y = V_{\text{avg}} t = \frac{gt^2}{2}$$

where V_{avg} = average velocity over period of time t. The equation for the path of the jet:

$$X^2 = C_v^2 4hY$$

ORIFICE DISCHARGE INTO DIVERGING CONICAL TUBES

This type of tube can greatly increase the flow through an orifice by reducing the pressure at the orifice below atmospheric. The formula that follows for the pressure at the entrance to the

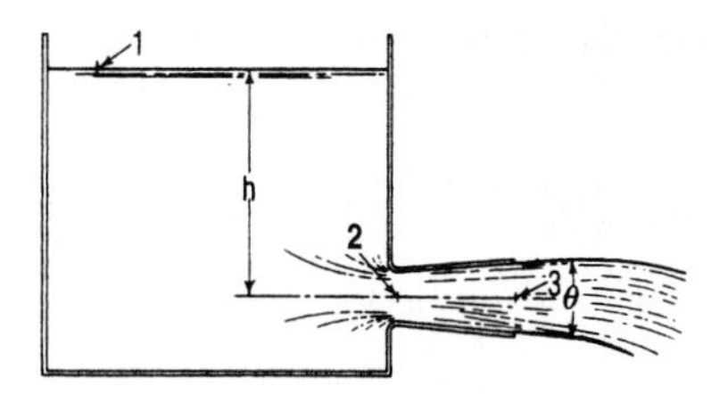

FIGURE 12.10 Diverging conical tube increases flow from a reservoir through an orifice by reducing the pressure below atmospheric.

tube is obtained by writing the Bernoulli equation for points 1 and 3 and points 1 and 2 in Fig. 12.10:

$$p_2 = wh\left[1 - \left(\frac{a_3}{a_2}\right)^2\right]$$

where p_2 = gage pressure at tube entrance, lb/ft² (Pa)

w = unit weight of water, lb/ft³ (kg/m³)

h = head on centerline of orifice, ft (m)

a_2 = area of smallest part of jet (vena contracta, if one exists), ft² (m)

a_3 = area of discharge end of tube, ft² (m²)

Discharge is also calculated by writing the Bernoulli equation for points 1 and 3 in Fig. 12.10.

For this analysis to be valid, the tube must flow full, and the pressure in the throat of the tube must not fall to the vapor pressure of water. Experiments by Venturi show the most efficient angle θ to be around 5°.

WATER HAMMER

Water hammer is a change is pressure, either above or below the normal pressure, caused by a variation of the flow rate in a pipe.

The equation for the velocity of a wave in a pipe is

$$U = \sqrt{\frac{E}{\rho}} \sqrt{\frac{1}{1 + ED/E_p t}}$$

where U = velocity of pressure wave along pipe, ft/s (m/s)

E = modulus of elasticity of water, 43.2×10^6 lb/ft^2 (2.07×10^6 kPa)

ρ = density of water, 1.94 lb·s/ft^4 (specific weight divided by acceleration due to gravity)

D = diameter of pipe, ft (m)

E_p = modulus of elasticity of pipe material, lb/ft^2 (kg/m^2)

t = thickness of pipe wall, ft (m)

PIPE STRESSES PERPENDICULAR TO THE LONGITUDINAL AXIS

The stresses acting perpendicular to the longitudinal axis of a pipe are caused by either internal or external pressures on the pipe walls.

Internal pressure creates a stress commonly called hoop tension. It may be calculated by taking a free-body diagram

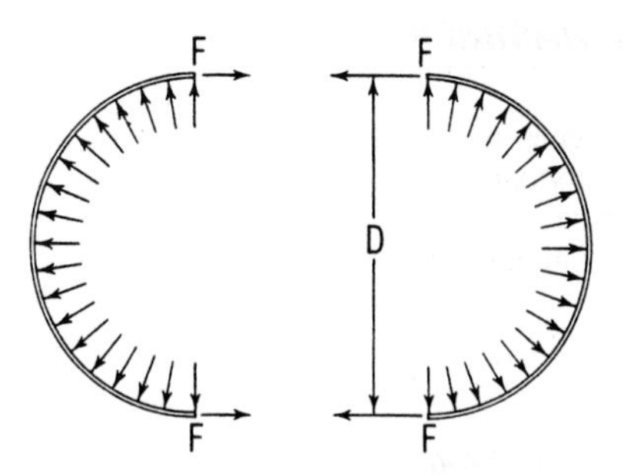

FIGURE 12.11 Internal pipe pressure produces hoop tension.

of a 1-in (25.4-mm)-long strip of pipe cut by a vertical plane through the longitudinal axis (Fig. 12.11). The forces in the vertical direction cancel out. The sum of the forces in the horizontal direction is

$$pD = 2F$$

where p = internal pressure, lb/in^2 (MPa)

D = outside diameter of pipe, in (mm)

F = force acting on each cut of edge of pipe, lb (N)

Hence, the stress, lb/in^2 (MPa) on the pipe material is

$$f = \frac{F}{A} = \frac{pD}{2t}$$

where A = area of cut edge of pipe, ft^2 (m^2); and t = thickness of pipe wall, in (mm).

TEMPERATURE EXPANSION OF PIPE

If a pipe is subject to a wide range of temperatures, the stress, lb/in^2 (MPa), due to a temperature change is

$$f = cE\,\Delta T$$

where E = modulus of elasticity of pipe material, lb/in^2 (MPa)

ΔT = temperature change from installation temperature

c = coefficient of thermal expansion of pipe material

The movement that should be allowed for, if expansion joints are to be used, is

$$\Delta L = Lc\,\Delta T$$

where ΔL = movement in length L of pipe, and L = length between expansion joints.

FORCES DUE TO PIPE BENDS

It is common practice to use thrust blocks in pipe bends to take the forces on the pipe caused by the momentum change and the unbalanced internal pressure of the water.

The force diagram in Fig. 12.12 is a convenient method for finding the resultant force on a bend. The forces can be resolved into X and Y components to find the magnitude and direction of the resultant force on the pipe. In Fig. 12.12,

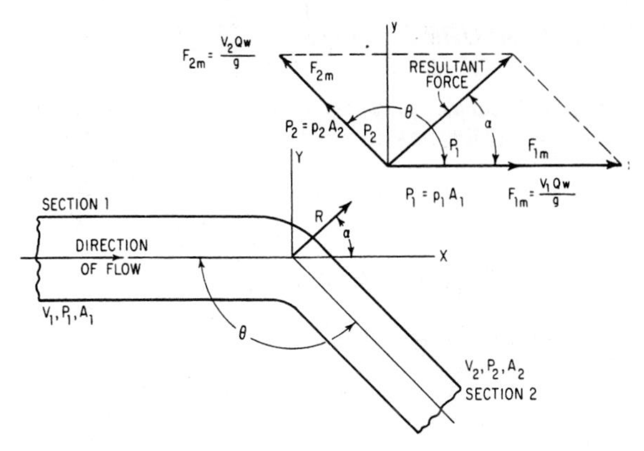

FIGURE 12.12 Forces produced by flow at a pipe bend and change in diameter.

V_1 = velocity before change in size of pipe, ft/s (m/s)

V_2 = velocity after change in size of pipe, ft/s (m/s)

p_1 = pressure before bend or size change in pipe, lb/ft^2 (kPa)

p_2 = pressure after bend or size change in pipe, lb/ft^2 (kPa)

A_1 = area before size change in pipe, ft^2 (m^2)

A_2 = area after size change in pipe, ft^2 (m^2)

F_{2m} = force due to momentum of water in section 2 = V_2Qw/g

F_{1m} = force due to momentum of water in section $1 = V_1Qw/g$

P_2 = pressure of water in section 2 times area of section $2 = p_2A_2$

P_1 = pressure of water in section 1 times area of section $1 = p_1A_1$

w = unit weight of liquid, lb/ft^3 (kg/m^3)

Q = discharge, ft^3/s (m^3/s)

If the pressure loss in the bend is neglected and there is no change in magnitude of velocity around the bend, a quick solution is

$$R = 2A\left(w\frac{V^2}{g} + p\right)\cos\frac{\theta}{2}$$

$$\alpha = \frac{\theta}{2}$$

where R = resultant force on bend, lb (N)

α = angle R makes with F_{1m}

p = pressure, lb/ft^2 (kPa)

w = unit weight of water, 62.4 lb/ft^3 (998.4 kg/m^3)

V = velocity of flow, ft/s (m/s)

g = acceleration due to gravity, 32.2 ft/s^2 (9.81 m/s^2)

A = area of pipe, ft^2 (m^2)

θ = angle between pipes ($0° \leq \theta \leq 180°$)

CULVERTS

A culvert is a closed conduit for the passage of surface drainage under a highway, a railroad, a canal, or other embankment. The slope of a culvert and its inlet and outlet conditions are usually determined by the topography of the site. Because of the many combinations obtained by varying the entrance conditions, exit conditions, and slope, no single formula can be given that applies to all culvert problems.

The basic method for determining discharge through a culvert requires application of the Bernoulli equation between a point just outside the entrance and a point somewhere downstream.

Entrance and Exit Submerged

When both the exit and entrance are submerged (Fig. 12.13), the culvert flows full, and the discharge is independent of the slope. This is normal pipe flow and is easily solved by using the Manning or Darcy–Weisbach formula for friction loss.

From the Bernoulli equation for the entrance and exit, and the Manning equation for friction loss, the following equation is obtained:

$$H = (1 - K_e)\frac{V^2}{2g} + \frac{V^2 n^2 L}{2.21 R^{4/3}}$$

Solution for the velocity of flow yields

$$V = \sqrt{\frac{H}{(1 + K_e/2g) + (n^2 L/2.21 R^{4/3})}}$$

where H = elevation difference between headwater and tailwater, ft (m)

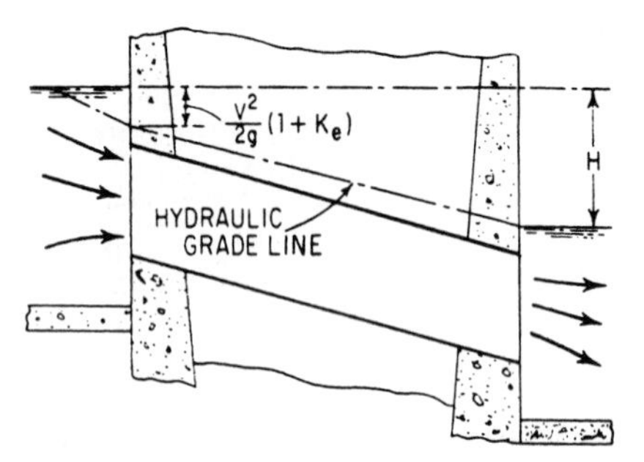

FIGURE 12.13 With entrance and exit of a culvert submerged, normal pipe flow occurs. Discharge is independent of slope. The fluid flows under pressure. Discharge may be determined from Bernoulli and Manning equations.

V = velocity in culvert, ft/s (m/s)

g = acceleration due to gravity, 32.2 ft/s^2 (9.81 m/s^2)

K_e = entrance-loss coefficient

n = Manning's roughness coefficient

L = length of culvert, ft (m)

R = hydraulic radius of culvert, ft (m)

The preceding equation can be solved directly because the velocity is the only unknown.

Culverts on Subcritical Slopes

Critical slope is the slope just sufficient to maintain flow at critical depth. When the slope is less than critical, the flow is considered subcritical.

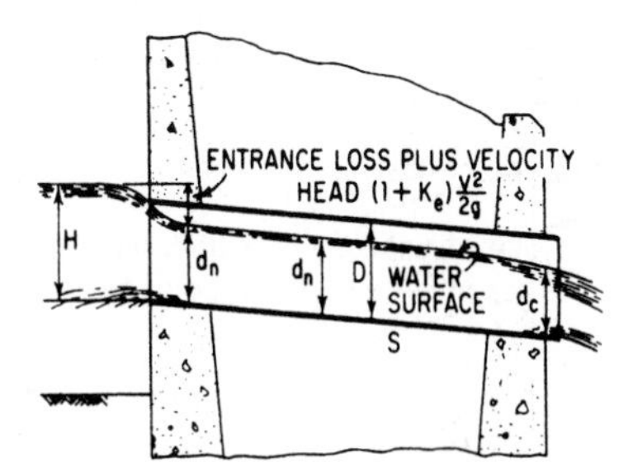

FIGURE 12.14 Open-channel flow occurs in a culvert with free discharge and normal depth d_n greater than the critical depth d_c when the entrance is unsubmerged or slightly submerged. Discharge depends on head H, loss at entrance, and slope of culvert.

Entrance Submerged or Unsubmerged but Free Exit. For these conditions, depending on the head, the flow can be either pressure or open channel (Fig. 12.14).

The discharge for the open-channel condition is obtained by writing the Bernoulli equation for a point just outside the entrance and a point a short distance downstream from the entrance. Thus,

$$H = K_e \frac{V^2}{2g} + \frac{V^2}{2g} + d_n$$

The velocity can be determined from the Manning equation:

$$V^2 = \frac{2.2SR^{4/3}}{n^2}$$

By substituting this into

$$H = (1 + K_e)\frac{2.2}{2gn^2} SR^{4/3} + d_n$$

where H = head on entrance measured from bottom of culvert, ft (m)

K_e = entrance-loss coefficient

S = slope of energy grade line, which for culverts is assumed to equal slope of bottom of culvert

R = hydraulic radius of culvert, ft (m)

d_n = normal depth of flow, ft (m)

To solve the preceding head equation, it is necessary to try different values of d_n and corresponding values of R until a value is found that satisfies the equation.

OPEN-CHANNEL FLOW

Free surface flow, or open-channel flow, includes all cases of flow in which the liquid surface is open to the atmosphere. Thus, flow in a pipe is open-channel flow if the pipe is only partly full.

A *uniform channel* is one of constant cross section. It has *uniform flow* if the grade, or slope, of the water surface is the same as that of the channel. Hence, depth of flow is constant throughout. *Steady flow* in a channel occurs if the depth at any location remains constant with time.

The *discharge* Q at any section is defined as the volume of water passing that section per unit of time. It is expressed in cubic feet per second, ft^3/s (cubic meter per second, m^3/s), and is given by

$$Q = VA$$

where V = average velocity, ft/s (m/s)

A = cross-sectional area of flow, ft^2 (m^2)

When the discharge is constant, the flow is said to be *continuous* and therefore

$$Q = V_1 A_1 = V_2 A_2 = \cdots$$

where the subscripts designate different channel sections. This preceding equation is known as the continuity equation for continuous steady flow.

Depth of flow d is taken as the vertical distance, ft (m), from the bottom of a channel to the water surface. The *wetted perimeter* is the length, ft (m), of a line bounding the cross-sectional area of flow minus the free surface width. The *hydraulic radius* R equals the area of flow divided by its wetted perimeter. The *average velocity* of flow V is defined as the discharge divided by the area of flow:

$$V = \frac{Q}{A}$$

The velocity head H_V, ft (m), is generally given by

$$H_V = \frac{V^2}{2g}$$

where V = average velocity, ft/s (m/s); and g = acceleration due to gravity, 32.2 ft/s^2 (9.81 m/s^2).

The *true velocity head* may be expressed as

$$H_{Va} = \alpha \frac{V^2}{2g}$$

where α is an empirical coefficient that represents the degree of turbulence. Experimental data indicate that α may vary from about 1.03 to 1.36 for prismatic channels. It is, however, normally taken as 1.00 for practical hydraulic work and is evaluated only for precise investigations of energy loss.

The total energy per pound (kilogram) of water relative to the bottom of the channel at a vertical section is called the *specific energy head* H_e. It is composed of the depth of flow at any point, plus the velocity head at the point. It is expressed in feet (meter) as

$$H_e = d + \frac{V^2}{2g}$$

A longitudinal profile of the elevation of the specific energy head is called the *energy grade line,* or the *total-head line* (Fig. 12.15). A longitudinal profile of the water surface is called the *hydraulic grade line.* The vertical distance between these profiles at any point equals the velocity head at that point.

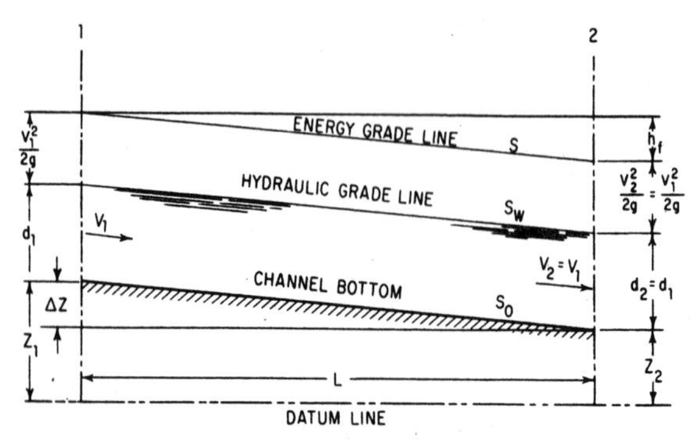

FIGURE 12.15 Characteristics of uniform open-channel flow.

Loss of head due to friction h_f in channel length L equals the drop in elevation of the channel ΔZ in the same distance.

Normal Depth of Flow

The depth of equilibrium flow that exists in the channel of Fig. 12.15 is called the normal depth d_n. This depth is unique for specific discharge and channel conditions. It may be computed by a trial-and-error process when the channel shape, slope, roughness, and discharge are known. A form of the Manning equation is suggested for this calculation:

$$AR^{2/3} = \frac{Qn}{1.486S^{1/2}}$$

where A = area of flow, ft^2 (m^2)

R = hydraulic radius, ft (m)

Q = amount of flow or discharge, ft^3/s (m^3/s)

n = Manning's roughness coefficient

S = slope of energy grade line or loss of head, ft (m), due to friction per linear ft (m), of channel

$AR^{2/3}$ is referred to as a *section factor.*

Critical Depth of Open-Channel Flow

For a given value of specific energy, the critical depth gives the greatest discharge, or conversely, for a given discharge, the specific energy is a minimum for the critical depth.

For rectangular channels, the critical depth, d_c ft (m), is given by

$$d_c = \sqrt[3]{\frac{Q^2}{b^2 g}}$$

where d_c = critical depth, ft (m)

Q = quantity of flow or discharge, ft^3/s (m^3/s)

b = width of channel, ft (m)

MANNING'S EQUATION FOR OPEN CHANNELS

One of the more popular of the numerous equations developed for determination of flow in an open channel is Manning's variation of the *Chezy formula*:

$$V = C\sqrt{RS}$$

where R = hydraulic radius, ft (m)

V = mean velocity of flow, ft/s (m/s)

S = slope of energy grade line or loss of head due to friction, ft/linear ft (m/m), of channel

C = Chezy roughness coefficient

Manning proposed:

$$C = \frac{1.486^{1/6}}{n}$$

where n is the coefficient of roughness in the Ganguillet–Kutter formula.

When Manning's C is used in the Chezy formula, the Manning equation for flow velocity in an open channel results:

$$V = \frac{1.486}{n} R^{2/3} S^{1/2}$$

Because the discharge $Q = VA$, this equation may be written:

$$Q = \frac{1.486}{n} AR^{2/3} S^{1/2}$$

where A = area of flow, ft^2 (m^2); and Q = quantity of flow, ft^3/s (m^3/s).

HYDRAULIC JUMP

This is an abrupt increase in depth of rapidly flowing water (Fig. 12.16). Flow at the jump changes from a supercritical to a subcritical stage with an accompanying loss of kinetic energy. Depth at the jump is not discontinuous. The change in depth occurs over a finite distance, known as the length of jump. The upstream surface of the jump, known as the roller, is a turbulent mass of water.

The depth before a jump is the *initial depth,* and the depth after a jump is the *sequent depth.* The specific energy for the sequent depth is less than that for the initial depth because of the energy dissipation within the jump. (Initial and sequent depths should not be confused with the depths of equal energy, or alternate depths.)

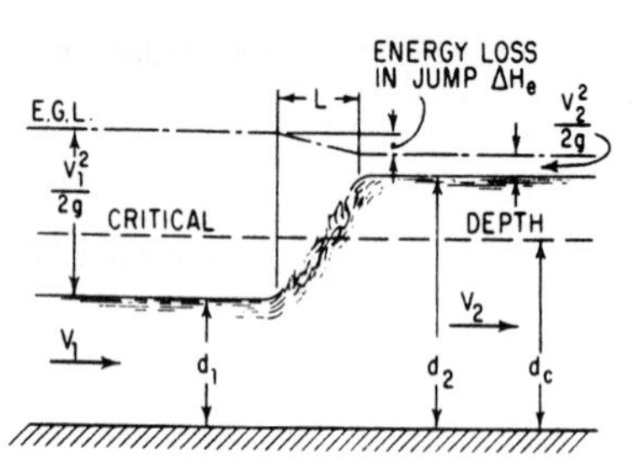

FIGURE 12.16 Hydraulic jump.

The pressure force F developed in hydraulic jump is

$$F = \frac{d_2^2 w}{2} - \frac{d_1^2 w}{2}$$

where d_1 = depth before jump, ft (m)

d_2 = depth after jump, ft (m)

w = unit weight of water, lb/ft³ (kg/m³)

The rate of change of momentum at the jump per foot width of channel equals

$$F = \frac{MV_1 - MV_2}{t} = \frac{qw}{g}(V_1 - V_2)$$

where M = mass of water, lb·s²/ft (kg·s²/m)

V_1 = velocity at depth d_1, ft/s (m/s)

V_2 = velocity at depth d_2, ft/s (m/s)

q = discharge per foot width of rectangular channel, ft³/s (m³/s)

t = unit of time, s

g = acceleration due to gravity, 32.2 ft/s² (9.81 kg/s²)

Then
$$V_1^2 = \frac{gd_2}{2d_1}(d_2 + d_1)$$

$$d_2 = \frac{-d_1}{2} + \sqrt{\frac{2V_1^2 d_1}{g} + \frac{d_1^2}{4}}$$

$$d_1 = \frac{-d_2}{2} + \sqrt{\frac{2V_2^2 d_2}{g} + \frac{d_2^2}{4}}$$

The head loss in a jump equals the difference in specific-energy head before and after the jump. This difference (Fig. 12.17) is given by

$$\Delta H_e = H_{e1} - H_{e2} = \frac{(d_2 - d_1)^3}{4d_1 d_2}$$

where H_{e1} = specific-energy head of stream before jump, ft (m); and H_{e2} = specific-energy head of stream after jump, ft (m).

The depths before and after a hydraulic jump may be related to the critical depth by

$$d_1 d_2 \frac{d_1 + d_2}{2} = \frac{q^2}{g} = d_c^3$$

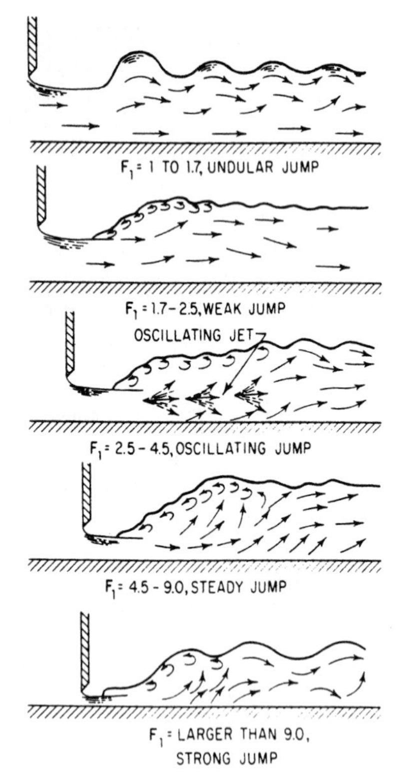

FIGURE 12.17 Type of hydraulic jump depends on Froude number.

where q = discharge, ft³/s (m³/s) per ft (m) of channel width; and d_c = critical depth for the channel, ft (m).

It may be seen from this equation that if $d_1 = d_c$, d_2 must also equal d_c.

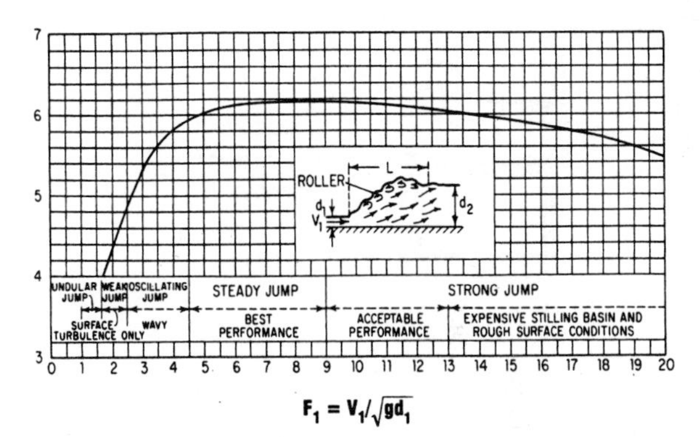

FIGURE 12.18 Length of hydraulic jump in a horizontal channel depends on sequent depth d_2 and the Froude number of the approaching flow.

Figure 12.18 shows how the length of hydraulic jump may be computed using the Froude number and the L/d_2 ratio.

NONUNIFORM FLOW IN OPEN CHANNELS

Symbols used in this section are V = velocity of flow in the open channel, ft/s (m/s); D_c = critical depth, ft (m); g = acceleration due to gravity, ft/s^2 (m/s^2); Q = flow rate, ft^3/s (m^3/s); q = flow rate per unit width, ft^3/ft (m^3/m); H_m = minimum specific energy, ft·lb/lb (kg·m/kg). Channel dimensions are in feet or meters and the symbols for them are given in the text and illustrations.

Nonuniform flow occurs in open channels with gradual or sudden changes in the cross-sectional area of the fluid stream. The terms *gradually varied flow* and *rapidly varied flow* are used to describe these two types of nonuniform

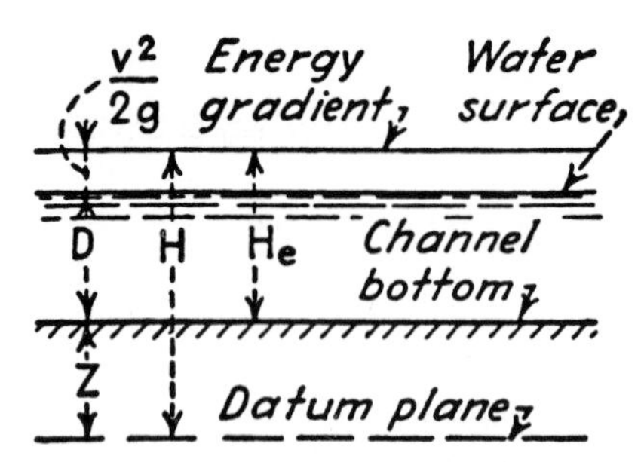

FIGURE 12.19 Energy of open-channel fluid flow.

flow. Equations are given next for flow in (1) rectangular cross-section channels, (2) triangular channels, (3) parabolic channels, (4) trapezoidal channels, and (5) circular channels. These five types of channels cover the majority of actual examples met in the field. Figure 12.19 shows the general energy relations in open-channel flow.

Rectangular Channels

In a rectangular channel, the critical depth D_c equals the mean depth D_m; the bottom width of the channel b equals the top width T; and when the discharge of fluid is taken as the flow per foot (meter) of width q of the channel, both b and T equal unity. Then V_c, the average velocity, is

$$V_c = \sqrt{gD_c} \tag{12.1}$$

and

$$D_c = \frac{V_c^2}{g} \tag{12.2}$$

Also $$Q = \sqrt{g}\ bD_c^{3/2} \tag{12.3}$$

where g = acceleration due to gravity in USCS or SI units.

$$q = \sqrt{g}\ D_c^{3/2} \tag{12.4}$$

and $$D_c = \sqrt[3]{\frac{q^2}{g}} \tag{12.5}$$

The minimum specific energy is

$$H_m = {}^3\!/_2 D_c \tag{12.6}$$

and the critical depth is

$$D_c = {}^2\!/_3 H_m \tag{12.7}$$

Then the discharge per foot (meter) of width is given by

$$q = \sqrt{g}\ ({}^2\!/_3)^{3/2} H_m^{3/2} \tag{12.8}$$

With $g = 32.16$, Eq. (12.8) becomes

$$q = 3.087 H_m^{3/2} \tag{12.9}$$

Triangular Channels

In a triangular channel (Fig. 12.20), the maximum depth D_c and the mean depth D_m equal $^1\!/_2\ D_c$. Then,

$$V_c = \sqrt{\frac{gD_c}{2}} \tag{12.10}$$

and
$$D_c = \frac{2V_c^2}{g} \tag{12.11}$$

As shown in Fig. 12.20, z is the slope of the channel sides, expressed as a ratio of horizontal to vertical; for symmetrical sections, $z = e/D_c$. The area, $a = zD_c^2$. Then,

$$Q = \sqrt{\frac{g}{2}}\, zD_c^{5/2} \tag{12.12}$$

With $g = 32.16$,

$$Q = 4.01zD_c^{5/2} \tag{12.13}$$

and
$$D_c = \sqrt[5]{\frac{2Q^2}{gz^2}} \tag{12.14}$$

or
$$Q = \sqrt{\frac{g}{2}}\left(\frac{4}{5}\right)^{5/2} zH_m^{5/2} \tag{12.15}$$

With $g = 32.16$,

$$Q = 2.295zH_m^{5/2} \tag{12.16}$$

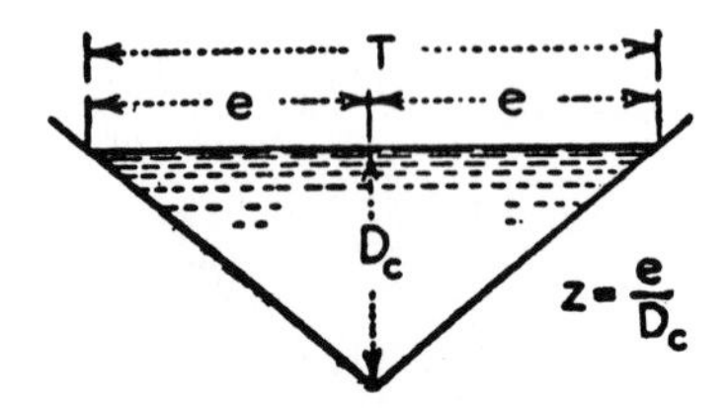

FIGURE 12.20 Triangular open channel.

Parabolic Channels

These channels can be conveniently defined in terms of the top width T and the depth D_c. Then the area $a = \frac{2}{3}D_cT$ and the mean depth $= D_m$.

Then (Fig. 12.21),

$$V_c = \sqrt{\tfrac{2}{3}gD_c} \tag{12.17}$$

and

$$D_c = \frac{3}{2}\frac{V_c^2}{g} \tag{12.18}$$

Further,

$$Q = \sqrt{\frac{8g}{27}}\,TD_c^{3/2} \tag{12.19}$$

With $g = 32.16$,

$$Q = 3.087TD_c^{3/2} \tag{12.20}$$

and

$$D_c = \frac{3}{2}\sqrt[3]{\frac{Q^2}{gT^2}} \tag{12.21}$$

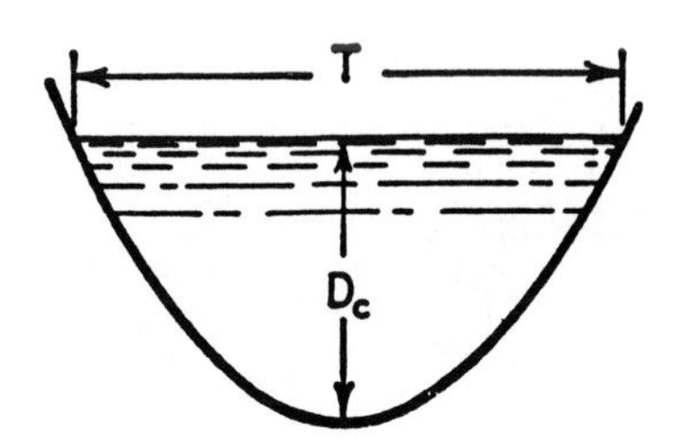

FIGURE 12.21 Parabolic open channel.

Also
$$Q = \sqrt{\frac{8g}{27}}\left(\frac{3}{4}\right)^{3/2} TH_m^{3/2} \tag{12.22}$$

With $g = 32.16$,

$$Q = 2.005TH_m^{3/2} \tag{12.23}$$

Trapezoidal Channels

Figure 12.22 shows a trapezoidal channel having a depth of D_c and a bottom width b. The slope of the sides, horizontal divided by vertical, is z. Expressing the mean depth D_m in terms of channel dimensions, the relations for critical depth D_c and average velocity V_c are

$$V_c = \sqrt{\frac{b + zD_c}{b + 2zD_c}\,gD_c} \tag{12.24}$$

and
$$D_c = \frac{V_c^2}{c} - \frac{b}{2z} + \sqrt{\frac{V_c^4}{g^2} + \frac{b^2}{4z^2}} \tag{12.25}$$

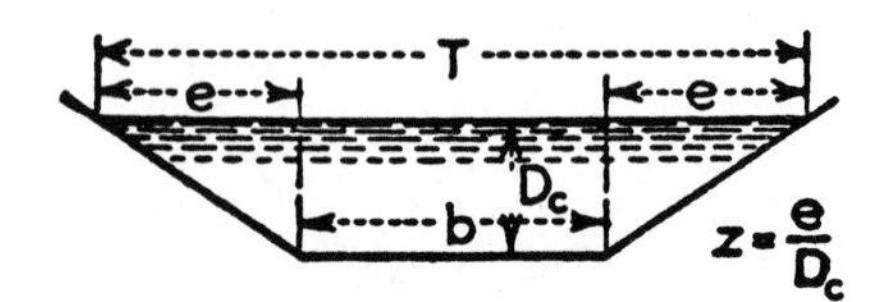

FIGURE 12.22 Trapezoidal open channel.

The discharge through the channel is then

$$Q = \sqrt{g \frac{(b + zD_c)^3}{b + 2zD_c}} D_c^{3/2} \tag{12.26}$$

Then, the minimum specific energy and critical depth are

$$H_m = \frac{3b + 5zD_c}{2b + 4zD_c} D_c \tag{12.27}$$

$$D_c = \frac{4zH_m - 3b + \sqrt{16z^2H_m^2 + 16zH_m b + 9b^2}}{10z} \tag{12.28}$$

Circular Channels

Figure 12.23 shows a typical circular channel in which the area a, top width T, and depth D_c are

$$a = \frac{d^2}{4}(\theta_r - \frac{1}{2}\sin 2\theta) \tag{12.29}$$

$$T = d \sin \theta \tag{12.30}$$

$$D_c = \frac{d}{2}(1 - \cos \theta) \tag{12.31}$$

Flow quantity is then given by

$$Q = \frac{2^{3/2}g^{1/2}(\theta_r - \frac{1}{2}\sin 2\theta)^{3/2}}{8(\sin \theta)^{1/2}(1 - \cos \theta)^{5/2}} D_c^{5/2} \tag{12.32}$$

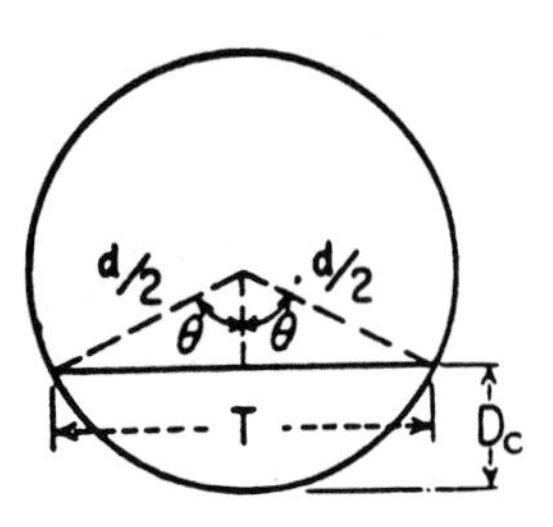

FIGURE 12.23 Circular channel.

WEIRS

A weir is a barrier in an open channel over which water flows. The edge or surface over which the water flows is called the *crest.* The overflowing sheet of water is the *nappe.*

If the nappe discharges into the air, the weir has *free discharge.* If the discharge is partly under water, the weir is *submerged* or drowned.

Types of Weirs

A weir with a sharp upstream corner or edge such that the water springs clear of the crest is a *sharp-crested weir* (Fig. 12.24). All other weirs are classed as *weirs not sharp crested.* Sharp-crested weirs are classified according to the shape of the weir opening, such as rectangular weirs, triangular or V-notch weirs, trapezoidal weirs, and parabolic weirs. Weirs not sharp crested are classified according to the shape of their cross section, such as broad-crested weirs, triangular weirs, and (as shown in Fig. 12.25) trapezoidal weirs.

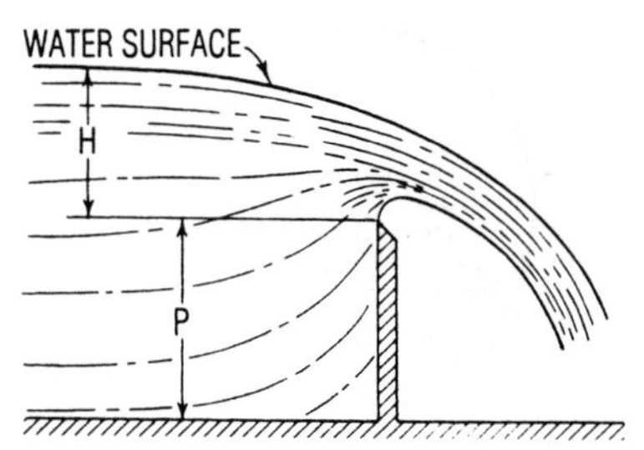

FIGURE 12.24 Sharp-crested weir.

The channel leading up to a weir is the *channel of approach.* The mean velocity in this channel is the *velocity of approach.* The depth of water producing the discharge is the *head.*

Sharp-crested weirs are useful only as a means of measuring flowing water. In contrast, weirs not sharp crested are commonly incorporated into hydraulic structures as control or regulation devices, with measurement of flow as their secondary function.

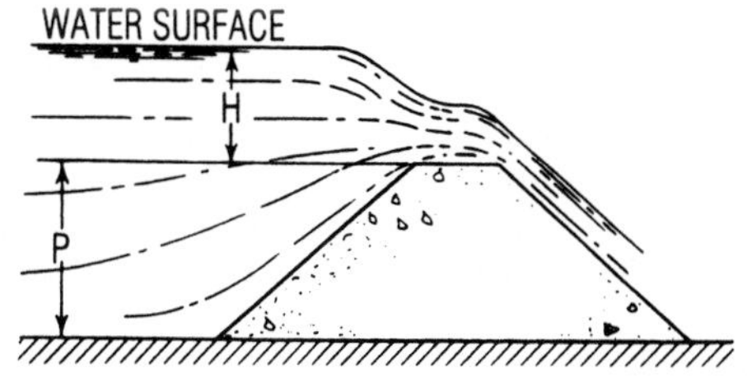

FIGURE 12.25 Weir not sharp crested.

FLOW OVER WEIRS

Rectangular Weir

The Francis formula for the discharge of a sharp-crested rectangular weir having a length b greater than $3h$ is

$$Q = 3.33\left(\frac{b - nh}{10}\right)[(h + h_0)^{3/2} - h_0^{3/2}]$$

where Q = discharge over weir, ft^3/s (m^3/s)

b = length of weir, ft (m)

h = vertical distance from level of crest of weir to water surface at point unaffected by weir drawdown (head on weir), ft (m)

n = number of end contractions (0, 1, or 2)

h_0 = head of velocity of approach [equal to $v_0^2/2g_c$, where v_0 = velocity of approach, (ft/s (m/s)], ft (m)

g_c = 32.2 (lb mass) (ft)/(lb force) (s^2)(m/s^2)

If the sides of the weir are coincident with the sides of the approach channel, the weir is considered to be suppressed, and $n = 0$. If both sides of the weir are far enough removed from the sides of the approach channel to permit free lateral approach of water, the weir is considered to be contracted, and $n = 2$. If one side is suppressed and one is contracted, $n = 1$.

TABLE 12.2 Discharge of Triangular Weirs

Notch (vertex) angle	Discharge formula[†]
90°	$Q = 0.685h^{2.45}$
60°	$Q = 1.45h^{2.47}$
30°	$Q = 2.49h^{2.48}$

[†] h is as defined above in the Francis formula.

Triangular Weir

The discharge of triangular weirs with notch angles of 30°, 60°, and 90° is given by the formulas in Table 12.2.

Trapezoidal (Cipolletti) Weir

The Cipolletti weir, extensively used for irrigation work, is trapezoidal in shape. The sides slope outward from the crest at an inclination of 1:4, (horizontal/vertical). The discharge is

$$Q = 3.367bh^{3/2}$$

where b, h, and Q are as defined earlier. The advantage of this type of weir is that no correction needs to be made for contractions.

Broad-Crested Weir

The discharge of a broad-crested weir is

$$Q = Cbh^{3/2}$$

TABLE 12.3 Variations in Head Ratio and Coefficient of Discharge for Broad-Crested Weirs

Ratio of actual head to design head	Coefficient of discharge
0.20	3.30
0.40	3.50
0.60	3.70
0.80	3.85
1.00	3.98
1.20	4.10
1.40	4.22

Values of C for broad-crested weirs with rounded upstream corners generally range from 2.6 to 2.9. For sharp upstream corners, C generally ranges from 2.4 to 2.6. Dam spillways are usually designed to fit the shape of the underside of a stream flowing over a sharp-crested weir. The coefficient C for such a spillway varies considerably with variation in the head, as shown in Table 12.3.

Q, b, and h are as defined for rectangular weirs.

PREDICTION OF SEDIMENT-DELIVERY RATE

Two methods of approach are available for predicting the rate of sediment accumulation in a reservoir; both involve predicting the rate of sediment delivery.

One approach depends on historical records of the silting rate for existing reservoirs and is purely empirical. The second general method of calculating the sediment-delivery rate involves determining the rate of sediment transport as a function of stream discharge and density of suspended silt.

The quantity of bed load is considered a constant function of the discharge because the sediment supply for the bed-load forces is always available in all but lined channels. An accepted formula for the quantity of sediment transported as bed load is the Schoklitsch formula:

$$G_b = \frac{86.7}{D_g^{1/2}} S^{3/2}(Q_i - bq_o)$$

where G_b = total bed load, lb/s (kg/s)

D_g = effective grain diameter, in (mm)

S = slope of energy gradient

Q_i = total instantaneous discharge, ft^3/s (m^3/s)

b = width of river, ft (m)

q_o = critical discharge, ft^3/s (m^3/s) per ft (m), of river width

$= (0.00532/S^{4/3})D_g$

An approximate solution for bed load by the Schoklitsch formula can be made by determining or assuming mean values of slope, discharge, and single grain size representative of the bed-load sediment. A mean grain size of 0.04 in (about 1 mm) in diameter is reasonable for a river with a slope of about 1.0 ft/mi (0.189 m/km).

EVAPORATION AND TRANSPIRATION

The *Meyer equation*, developed from Dalton's law, is one of many evaporation formulas and is popular for making evaporation-rate calculations:

$$E = C\,(e_w - e_a)\psi$$

$$\psi = 1 + 0.1w$$

where E = evaporation rate, in 30-day month

C = empirical coefficient, equal to 15 for small, shallow pools and 11 for large, deep reservoirs

e_w = saturation vapor pressure, in (mm), of mercury, corresponding to monthly mean air temperature observed at nearby stations for small bodies of shallow water or corresponding to water temperature instead of air temperature for large bodies of deep water

e_a = actual vapor pressure, in (mm), of mercury, in air based on monthly mean air temperature and relative humidity at nearby stations for small bodies of shallow water or based on information obtained about 30 ft (9.14 m) above the water surface for large bodies of deep water

w = monthly mean wind velocity, mi/h (km/h), at about 30 ft (9.14 m) aboveground

ψ = wind factor

As an example of the evaporation that may occur from a large reservoir, the mean annual evaporation from Lake Mead is 6 ft (1.82 m).

METHOD FOR DETERMINING RUNOFF FOR MINOR HYDRAULIC STRUCTURES

The most common means for determining runoff for minor hydraulic structures is the *rational formula*:

$$Q = CIA$$

where Q = peak discharge, ft^3/s (m^3/s)

C = runoff coefficient = percentage of rain that appears as direct runoff

I = rainfall intensity, in/h (mm/h)

A = drainage area, acres (m^2)

COMPUTING RAINFALL INTENSITY

Chow lists 24 rainfall-intensity formulas of the form:

$$I = \frac{KF^{n1}}{(t + b)^n}$$

where I = rainfall intensity, in/h (mm/h)

K, b, n, and n_1 = coefficient, factor, and exponents, respectively, depending on conditions that affect rainfall intensity

F = frequency of occurrence of rainfall, years

t = duration of storm, min

= time of concentration

Perhaps the most useful of these formulas is the *Steel formula*:

$$I = \frac{K}{t + b}$$

where K and b are dependent on the storm frequency and region of the United States (Fig. 12.26 and Table 12.4).

The Steel formula gives the average maximum precipitation rates for durations up to 2 h.

FIGURE 12.26 Regions of the United States for use with the Steel formula.

TABLE 12.4 Coefficients for Steel Formula

Frequency, years	Coefficients	Region						
		1	2	3	4	5	6	7
2	K	206	140	106	70	70	68	32
	b	30	21	17	13	16	14	11
4	K	247	190	131	97	81	75	48
	b	29	25	19	16	13	12	12
10	K	300	230	170	111	111	122	60
	b	36	29	23	16	17	23	13
25	K	327	260	230	170	130	155	67
	b	33	32	30	27	17	26	10

GROUNDWATER

Groundwater is subsurface water in porous strata within a zone of saturation. It supplies about 20 percent of the United States water demand.

Aquifers are groundwater formations capable of furnishing an economical water supply. Those formations from which extractions cannot be made economically are called *aquicludes*.

Permeability indicates the ease with which water moves through a soil and determines whether a groundwater formation is an aquifer or aquiclude.

The rate of movement of groundwater is given by Darcy's law:

$$Q = KIA$$

where Q = flow rate, gal/day (m^3/day)

K = hydraulic conductivity, ft/day (m/day)

I = hydraulic gradient, ft/ft (m/m)

A = cross-sectional area, perpendicular to direction of flow, ft^2 (m^2)

WATER FLOW FOR FIREFIGHTING

The total quantity of water used for fighting fires is normally quite small, but the demand rate is high. The fire demand as established by the American Insurance Association is

$$G = 1020\sqrt{P}(1 - 0.01\sqrt{P})$$

where G = fire-demand rate, gal/min (liter/s); and P = population, thousands.

FLOW FROM WELLS

The steady flow rate Q can be found for a gravity well by using the *Dupuit formula*:

$$Q = \frac{1.36K(H^2 - h^2)}{\log(D/d)}$$

where Q = flow, gal/day (liter/day)

K = hydraulic conductivity, ft/day (m/day), under 1:1 hydraulic gradient

H = total depth of water from bottom of well to free-water surface before pumping, ft (m)

h = H minus drawdown, ft (m)

D = diameter of circle of influence, ft (m)

d = diameter of well, ft (m)

The steady flow, gal/day (liter/day), from an artesian well is given by

$$Q = \frac{2.73Kt(H - h)}{\log(D/d)}$$

where t is the thickness of confined aquifer, ft (m).

ECONOMICAL SIZING OF DISTRIBUTION PIPING

An equation for the most economical pipe diameter for a distribution system for water is

$$D = 0.215\left(\frac{fbQ_a^3 S}{aiH_a}\right)^{1/7}$$

where D = pipe diameter, ft (m)

f = Darcy–Weisbach friction factor

b = value of power, \$/hp per year (\$/kW per year)

Q_a = average discharge, ft^3/s (m^3/s)

S = allowable unit stress in pipe, lb/in^2 (MPa)

a = in-place cost of pipe, \$/lb (\$/kg)

i = yearly fixed charges for pipeline (expressed as a fraction of total capital cost)

H_a = average head on pipe, ft (m)

VENTURI METER FLOW COMPUTATION

Flow through a venturi meter (Fig. 12.27) is given by

$$Q = cKd_2^2\sqrt{h_1 - h_2}$$

$$K = \frac{4}{\pi}\sqrt{\frac{2g}{1-(d_2/d_1)^2}}$$

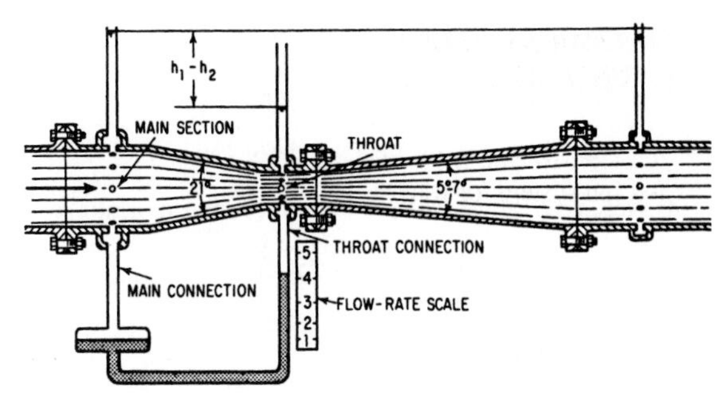

FIGURE 12.27 Standard venturi meter.

where Q = flow rate, ft^3/s (m^3/s)

c = empirical discharge coefficient dependent on throat velocity and diameter

d_1 = diameter of main section, ft (m)

d_2 = diameter of throat, ft (m)

h_1 = pressure in main section, ft (m) of water

h_2 = pressure in throat section, ft (m) of water

HYDROELECTRIC POWER GENERATION

Hydroelectric power is electrical power obtained from conversion of potential and kinetic energy of water. The potential energy of a volume of water is the product of its weight and the vertical distance it can fall:

$$PE = WZ$$

where PE = potential energy

W = total weight of the water

Z = vertical distance water can fall

Power is the rate at which energy is produced or utilized:

$$1 \text{ horsepower (hp)} = 550 \text{ ft}\cdot\text{lb/s}$$

$$1 \text{ kilowatt (kW)} = 738 \text{ ft}\cdot\text{lb/s}$$

$$1 \text{ hp} = 0.746 \text{ kW}$$

$$1 \text{ kW} = 1.341 \text{ hp}$$

Power obtained from water flow may be computed from

$$\text{hp} = \frac{\eta Q w h}{550} = \frac{\eta Q h}{8.8}$$

$$\text{kW} = \frac{\eta Q w h}{738} = \frac{\eta Q h}{11.8}$$

where kW = kilowatt

hp = horsepower

Q = flow rate, ft^3/s (m^3/s)

w = unit weight of water = 62.4 lb/ft^3 (998.4 kg/m^3)

h = effective head = total elevation difference minus line losses due to friction and turbulence, ft (m)

η = efficiency of turbine and generator

INDEX

Adjustment factors for lumber, 224–233
Allowable-stress design, 285–297

Beam formulas, 16–98
  beam formulas and elastic diagrams, 29–39
  beams of uniform strength, 63
  characteristics of loadings, 52
  combined axial and bending loads, 92
  continuous beams, 16, 27, 51
  curved beams, 82–88
    eccentrically curved, 86
  eccentric loading, 94–96
  elastic-curve equations for prismatic beams, 40–51
  elastic lateral buckling, 88–92
  natural circular and periods of vibration, 96–98
  rolling and moving loads, 79–82
  safe loads for beams of various types, 64–78
    parabolic beam, 64
    steel beam, 66
    triangular beam, 65
  ultimate strength of continuous beams, 53–62
    Castigliano's theorem, 62
    Maxwell's theorem, 62
  unsymmetrical bending, 93
Blasting, vibration control in, 280–282
Bridge and suspension-cable formulas, 322–354
  allowable-stress design, 323

Bridge and suspension-cable formulas (*Continued*)
for bridge beams, 325–327
for bridge columns, 323
for shear, 339–340
bearing on milled surfaces, 332
bridge fasteners, 333
cable systems, 353
composite construction in highway bridges, 333
bending stresses, 335
effective width of slabs, 334
shear range, 335–337
span/depth ratios, 334
general relations for suspension cables, 341–352
keeping strength at different levels, 348
parabola, 347
supports at different levels, 348
supports at same level, 349–352
hybrid bridge girders, 329
load-and-resistance factor design, 324–331
for bridge beams, 330–331
for bridge columns, 324–325
maximum width/thickness ratios, 341

Bridge and suspension-cable formulas (*Continued*)
number of connectors in bridges, 337–339
ultimate strength of connectors, 339
shear strength design for bridges, 322–323
stiffeners on bridge girders, 327
longitudinal stiffeners, 328
suspension cables, 341–352
catenary cable sag, 344
parabolic cable tension and length, 341–344
Building and structures formulas, 284–319
allowable-stress design, 285–297
for building beams, 287–290
for building columns, 285
for shear in buildings, 295–297
bearing plates, 298–300
bents and shear in walls, 304–306
deflections of, 304
column base plates, 300
combined axial compression or tension and bending, 306–307
composite construction, 313–315

Building and structures formulas (*Continued*)
design of stiffeners under load, 311–312
fasteners in buildings, 312–313
load-and-resistance factor design, 284–294
for building beams, 290–294
for columns, 287
for shear, 284–285
milled surfaces, bearing on, 301
number of connections required, 316–318
shear on connectors, 317
plate girders in buildings, 302–304
ponding considerations, 318–319
stresses in thin shells, 297
webs under concentrated loads, 308–310

Cable systems, 353
California bearing ratio, 274
Cantilever retaining walls, 208–211
Chezy formula, 399
Circular channels, 435
Circular curves, 356–359
Column formulas, 100–130
Columns, lumber, 218–220
Commonly used USCS and SI units, 3
Composite construction, 313–315, 333–337
Concrete, formulas for, 148–212
braced and unbraced frames, 201–202
cantilever retaining walls, 208–211
column moments, 199–200
compression development lengths, 170
continuous beams, 16, 27, 52, 151, 153–162
one-way slabs, 151–153
crack control, 170
deflection, computation for, 172–173
design methods, 153–162
beams, 153–162
columns, 162–167
flat-plate construction, 195–197
direct design method, 195–197
flat-slab construction, 192–195
gravity retaining walls, 205–208
hardened-state properties, 167–168
job mix volume, 148
load-bearing walls, 202–203
modulus of elasticity, 150

Concrete, formulas for (*Continued*)
properties in hardened state, 167–168
reinforced, 148–212
required strength, 171
shear in slabs, 197–199
shear walls, 203–205
spirals, 200–201
tensile strength, 151
tension development length, 169
ultimate-strength design of I and T beams, 186–187
ultimate-strength design of rectangular beams, 173–174
balanced reinforcing, 174
with compression bars, 183–185
development of tensile reinforcement, 177
hooks on bars, 178
moment capacity, 175
ultimate-strength design for torsion, 189–191
wall footings, 211–212
water-cementation ratio, 148
working-stress design:
for allowable bending moment, 179
for allowable shear, 180–181
of I and T beams, 187–189

Concrete, formulas for (*Continued*)
of rectangular beams, 183–185
for torsion, 189–191
Continuous beams, 16, 27, 51, 153–162
Conversion factors for civil engineering, 2–14
Conversion table, typical, 4
Crack control, 170
Culverts, highway, designing, 371–374

Darcy–Weisbach formula, 398
Design methods, 153–167
beams, 153–162
columns, 162–167

Earthmoving formulas, 276–278
Elastic–curve equations for prismatic beams, 40–51
Expansion, temperature, of pipe, 414

Factors, adjustment, for lumber, 224–233
conversion, 2–14
Fasteners, for lumber, 233–236
Fixed-end moments, 52
Flat-plate construction, 195–197

Flat-slab construction, 192–195
Forest Products Laboratory, 221

Geometric properties of sections, 17–28
Grading of lumber, 214

Highway and road formulas, 356–379
  American Association of State Highway and Transportation Officials (AASHTO), 363–365
  circular curves, 356–359
    equations of, 358
  culverts, highway, designing, 371–374
    American Iron and Steel Institute (AISI) design procedure, 374–379
    allowable wall stress, 376
    bolted seams, checking of, 379
    handling stiffness, 378
    ring compression, 376
    wall thickness, 378
  highway alignments, 362
    curves and driver safety, 361
    stopping sight distance, 363–365
    stationing, 362

Highway and road formulas (*Continued*)
  interchanges, types of, 367
  parabolic curves, 359
    equations of, 360
  street and maneuvering space, 370
  structural numbers for flexible pavements, 368–370
  transition (special) curves, 370
  turning lanes, 368–369
Hydraulic and waterworks formulas, 382–450
  capillary action, 382–386
  computing rainfall intensity, 443–445
  culverts, 417
    entrance and exit submerged, 417
    on subcritical slopes, 418–420
  economical sizing of distribution piping, 448
  evaporation and transpiration, 442
  flow from wells, 447
  flow over weirs, 438
    broad-crested weir, 439
    rectangular weir, 438
    trapezoidal (Cipolletti) weir, 439
    triangular weir, 439

Hydraulic and waterworks formulas (*Continued*)
- flow through orifices, 406–409
  - discharge under falling head, 409
  - submerged orifices, 408
- fluid flow in pipes, 395–403
  - Chezy formula, 399
  - Darcy–Weisbach formula, 398
  - Hazen-Williams formula, 401
  - Manning's formula, 401
  - turbulent flow, 397
- fluid jets, 409
- forces due to pipe bends, 414–416
- fundamentals of fluid flow, 388–392
- groundwater, 446
- hydraulic jump, 425–429
- hydroelectric power generation, 449–450
- Manning's equation for open channels, 424
- method for determining runoff, 443
- nonuniform flow in open channels, 429–435
  - circular channels, 435
  - parabolic channels, 433
  - rectangular channels, 430
  - trapezoidal channels, 434
  - triangular channels, 431

Hydraulic and waterworks formulas (*Continued*)
- open-channel flow, 420–423
  - critical depth of flow, 423
  - normal depth of flow, 423
- orifice discharge, 410
- pipe stresses, 412–413
- prediction of sediment delivery rate, 440
- pressure changes caused by pipe size changes, 403–404
  - bends and standard fitting losses, 405
  - gradual enlargements, 404
  - sudden contraction, 404
- similitude for physical models, 392–395
- submerged curved surfaces, pressure on, 387–388
- temperature expansion of pipe, 414
- venturi meter flow computation, 448
- viscosity, 386
- water flow for firefighting, 446
- water hammer, 412
- weirs, 436–439
  - types of, 436–437

Load-and-resistance factor design, 284–294, 324–331

Load-and-resistance factor design (*Continued*)
  for bridge beams, 330–331
  for bridge columns, 324–325
  for building beams, 290–294
  for building columns, 287
  for building shear, 284–285
Load-bearing walls, 202–203
Lumber formulas, 214–241
  adjustment factors for design values, 224–233
  beams, 215–218
  bearing area, 229
  bending and axial compression, 240
  bending and axial tension, 239
  column stability and buckling, 230–233
    for connections with fasteners, 236–238
  columns, 218–220
    in combined bending and axial load, 220
  compression, at angle to grain, 220
    on oblique plane, 223–224
  fasteners for lumber, 233
    nails, 233
    screws, 234–236
    spikes, 233
  Forest Products Laboratory recommendations, 221–222
  grading of lumber, 214
Lumber formulas (*Continued*)
  radial stresses and curvature, 236–238
  size and volume of lumber, 227

Manning's formula, 401
Maxwell's theorem, 62
Modulus of elasticity, 150

Open channels, nonuniform flow in, 429–435
Orifice discharge, 410
Orifices, flow through, 406–409

Parabolic channels, 433
Piles and piling formulas, 132–146
  allowable load, 132
  axial-load capacity, single piles, 143
  foundation-stability and, 139–143
  groups of piles, 136–139
  laterally loaded, 133
  shaft resistance, 145
  shaft settlement, 144
  toe capacity load, 134–135
Pipe bends, 414–416
Pipes, flow in, 395–403

Piping, economical sizing of, 448
Plate girders in buildings, 302–304
Ponding, roof, 318–319
Prismatic beams, elastic-curve equations for, 40–51

Rainfall intensity, computing, 443–445
Rectangular channels, 430
Retaining walls, forces on, 265
Road formulas, 356–379
Rolling and moving loads, 79–82
Roof slope to avoid ponding, 238–239

Safe loads for beams, 64–78
Sections, geometric properties of, 17–28
Sediment, prediction of delivery rate, 440
Similitude, physical models, 392–395
Sizes of lumber, 214–215
Soils and earthwork formulas, 258–282
- bearing capacity of, 270
- California bearing ratio, 274
- cohesionless soils, lateral pressure in, 266

Soils and earthwork formulas (*Continued*)
- cohesive soils, lateral pressure in, 267
- compaction equipment, 275
- compaction tests, 272
  - load-bearing, 273
- earthmoving formulas, 276–278
  - quantities hauled, 278
- forces on retaining walls, 265
- index parameters, 259–260
- internal friction and cohesion, 263
- lateral pressures, 264
- permeability, 274
- physical properties of soils, 258
- scraper production, 278
  - equipment required, 278
- settlement under foundations, 271
- stability of slopes, 269
  - cohesionless soils, 269
  - cohesive soils, 269
- surcharge lateral pressure, 268
- vertical pressures, 264
- vibration control in blasting, 280–282
- water pressure and soils, 268
- weights and volumes, relationships of, 261–262

Strength, tensile, 151
Submerged curved surfaces, 387–388
Surveying formulas, 244–256
- distance measurement with tapes, 247–250
- orthometric correction, 251–252
- photogrammetry, 255–256
- slope corrections, 250
- stadia surveying, 253–255
- temperature corrections, 250
- theory of errors, 245–247
- units of measurement, 244
- vertical control, 253

Suspension cables, 341–352

Table, conversion, 4
Temperature expansion of pipe, 414
Thin shells, 297
Timber engineering, 214–241
Trapezoidal channels, 434
Triangular channels, 431

Ultimate strength, of continuous beams, 53–62
Uniform strength, of beams, 63

Venturi meter flow computations, 448
Vibration, natural circular and periods of, 96–98
Viscosity, 386

Walls, load bearing, 202–203
Water flow for firefighting, 446
Weirs, 436–439
Wells, flow from, 447

ABOUT THE AUTHOR

Tyler G. Hicks, P.E., is a consulting engineer and a successful engineering book author. He has been involved in plant design and operation in a multitude of industries, has taught at several engineering schools, and has lectured both in the United States and abroad. Mr. Hicks holds a Bachelor's Degree in Mechanical Engineering from Cooper Union School of Engineering in New York.